T0093408

Blockchain Technology for IoT and Wireless Communications

Smart communications are the concept in which smart appliances and devices are integrated into an application that runs in a smart hand-held device. The residents of a smart home can have complete control over their home's electronic gadgets using wireless communications. Thus, it can help people control gadgets in the home/office remotely and often simultaneously, which increases convenience and reduces time spent on these tasks. The problem with security systems associated with smart devices is that they may be broken when there are loopholes or human mistakes. When security credentials are lost, overall security can also be lost. This is because smart technology is made up of plenty of devices that are integrated with Internet of Things (IoT) technology and the cloud. This environment can introduce many security issues, as discussed in this text. Blockchain is a promising technology that operates in a decentralized environment to protect devices and the data collected by devices from security and privacy issues by using wireless communication technology. Blockchain-enabled IoT can be used to achieve end-to-end security. Blockchain technology is already used in wireless sensor/communication networks to estimate and predict house data and civil structures.

IoT-integrated innovative applications like smart homes present unique security and privacy challenges. Scalability is the main problem as the current centralized IoT platforms have message routing mechanisms that create a bottleneck in scaling up too many devices used in IoT. As many devices are participating in generating data, such a setup may be subjected to Distributed Denial of Service (DDoS) attacks. Lack of data standards is another cause of concern as it leads to interoperability problems. Blockchain technology for IoT and wireless communications offers a promising solution for smart devices. These technologies can provide end-to-end security and overcome the aforementioned problems. The usage of open-standard distributed IoT solutions can solve many problems that are associated with centralized approaches. Blockchain technology is nothing but a distributed ledger of transactions. It offers direct communication to connected devices. Such devices collect data, and all legitimate participants can access said data. Thus, decentralized blockchain networks can provide improved security for IoT-based solutions.

Blockchain Technology for IoT and Wireless Communications

Edited by
Gajula Ramesh
Budati Anil Kumar
Praveen Jugge
Kolalapudi Lakshmi Prasad
Mohammad Kamrul Hasan

CRC Press is an imprint of the
Taylor & Francis Group, an informa business

Cover image: ©Shutterstock

First edition published 2024
by CRC Press
6000 Broken Sound Parkway NW, Suite 300, Boca Raton, FL 33487-2742

and by CRC Press
4 Park Square, Milton Park, Abingdon, Oxon, OX14 4RN

CRC Press is an imprint of Taylor & Francis Group, LLC

Library of Congress Cataloging-in-Publication Data
Names: Ramesh, Gajula, editor.
Title: Blockchain technology for IoT and wireless communications / edited by Gajula Ramesh, Budati Anil Kumar, Juggi Praveen, Kolalapudi Lakshmi Prasad, Mohammad Kamrul Hasan.
Description: First edition. | Boca Raton : CRC Press, 2023. | Includes bibliographical references and index.
Identifiers: LCCN 2023003309 (print) | LCCN 2023003310 (ebook) |
ISBN 9781032217840 (hbk) | ISBN 9781032217857 (pbk) | ISBN 9781003269991 (ebk)
Subjects: LCSH: Internet of things. | Wireless communication systems. | Blockchains (Databases)
Classification: LCC TK5105.8857 .B583 2023 (print) | LCC TK5105.8857 (ebook) | DDC 005.8--dc23/eng/20230215
LC record available at https://lccn.loc.gov/2023003309
LC ebook record available at https://lccn.loc.gov/2023003310

ISBN: 978-1-032-21784-0 (hbk)
ISBN: 978-1-032-21785-7 (pbk)
ISBN: 978-1-003-26999-1 (ebk)

DOI: 10.1201/9781003269991

Typeset in Caslon
by MPS Limited, Dehradun

Contents

About the Editors

Dr. G. Ramesh received a PhD degree in computer science and engineering from JNT University Anantapur, Ananthapuaramu, Andhra Pradesh in 2018, and a M.Tech. degree from the JNTUA College of Engineering, Ananthapuramu, in 2009. He is currently working as an associate professor at CSE Department, Gokaraju Rangaraju Institute of Engineering & Technology (Autonomous), Hyderabad- 500090, Telangana State, India. He has more than 11 years of experience in teaching and 6 years of experience in research. Additionally, he has published 35 research articles in leading journals (SCI, SCOPUS &DBLP), conference proceedings, 1 textbook and 3 international book chapters. He has 78 citations with an h-index of 4 for his research publications. He is a reviewer for reputed international conferences. He has organized and attended various FDPs and conferences/workshops. He also worked as coordinator for conducting the course "Database Application Software and Database Design", for UG and PG students of JNTUA CE, Ananthapuramu, in collaboration with Electronic and ICT Academy, NIT Warangal, and IIT Madras, Chennai. His current research interests include software engineering, machine learning, image processing, and big data analytics.

Dr. Budati Anil Kumar received his PhD degree in Electronics and Communication Engineering from GITAM Deemed to be

University, Hyderabad, in 2019, MTech degree from the LBRCE (JNTUK Kakinada), Mylavaram, in 2010, and BTech degree from VYCET (JNTU Hyderabad), Chirala, in 2007. He is currently working as an Associate Professor in the ECE Department, Koneru Lakshmaiah Education Foundation, Hyderabad, Telangana State, India. He has more than 13 years of experience in teaching and six years of experience in research. Additionally, he has published 50 research articles in highly reputed publisher journals and conferences that are indexed in SCIE, SCOPUS, etc. He acts as Guest Editor for *Peer-to-Peer Networking Journal* (Springer Publisher); *Cluster Computing Journal* (Springer Publisher); *IJPCC* (Emerald Publisher); *IET CAAI Transactions on Intelligence Technology, IET Networks, SWCC Journal* (Bentham Science Publisher); *IJSCC* (Inderscience Publisher); *IJUWBCS* (Inderscience Publisher) Journals. He has authored/edited textbooks entitled *Cognitive Radio: Computing Techniques, Network Security and Challenges* by Taylor and Francis publisher, UK, and *Cognitive Computing Models in Communication Systems* by Wiley & Scrivener Publishing, UK. He is a Senior Member of IEEE, Member of IEI, and Fellow of IETE since 2019 and has served as a reviewer for various international journals and conferences. He has conducted various FDPs as a coordinator, which is funded by AICTE New Delhi, DST New Delhi, and other agencies. He has attended various FDPs and Conferences/workshops in India and abroad and also acted as a speaker/session chair in various FDPs and Conferences. He has also served as a supervisor for PhD research scholars under multiple Universities in India. His current research interests include cognitive radio networks, software defined radio networks, artificial intelligence, 6G emerging technologies, mulsemedia computing, and UAVs in 5G and 6G.

Dr. J. Praveen received his PhD from Osmania University in power electronics, Hyderabad. His research was carried out at BHEL Research and Development Center with support of the University Grants Commission (UGC) junior research fellowship (JRF). He is contributing to a major research area in power electronics and has published more than 80 research papers in reputed international and national journals and conferences. He is a registered PhD guide in the Department of Electrical and Electronics Engineering, J.N.T

University, Hyderabad. He is presently guiding eight scholars of PhDs. He has international certification on "High Impact Teaching Skills" by Dale Carnegie & Associates Inc. Trainers (USA), Mission 10x, Wipro Technologies. He has Cambridge International Certification for Teachers and Trainers (CICTT) with Distinction. He is a senior member of IEEE, MIE – Life Member, ISTE. He has an Honor Code Certificate in course Circuits and Electronics 6.002X by Massachusetts Institute of Technology, USA (MIT), with 'A' Grade through edx. He received a best teacher award and best project award from ISTE. He is a certified labview associate developer from NI-National Instrument, USA.

Dr. K. Lakshmi Prasad received a PhD in water resources engineering from IIT Bombay (2000), M. Tech in hydraulic engineering from IIT Bombay (1985), and B. Tech in civil engineering from JNTU College of Engineering, Anantapur, AP (1983). He has handled two sponsored research and development projects worth 3.8 crores sponsored by the Department of Science & Technology, MHRD, Government of India 2014–18 as principal investigator. He has completed 33 years of service in teaching and research and published 27 papers in international refereed journals and international and national conferences. He is a fellow of the Institution of Engineers (India)- FIE, Member of Indian Society for Technical Education-MISTE. He served as principal at the GITAM Institute of Technology, GITAM Deemed to be University, Visakhapatnam 2012–2019 and served as a member, Board of Management, GITAM Deemed to be University, Visakhapatnam 2012–2019. He served as a member, Academic Council, GITAM Deemed to be University, Visakhapatnam 2012–2019 and served as dean, GITAM Institute of Technology and introduced the choice-based credit system with outcome-based educational objectives for UG and PG courses of engineering in 2015–16. He served as chairman, TEQIP-II (1.2), a world bank funded project of MHRD, Government of India during 2012–2017 and served as coordinator for TEQIP-Phase I, Bapatla Engineering College, Bapatla, A.P., during 2003–2009. He served as a member, Academic Council, Bapatla Engineering College, Bapatla, AP during 2011–17 and served as principal of BVSR Engineering College, Chimakurthy, Prakasam Dt.AP during 2010–12. He served as a principal, Bapatla Engineering College, Bapatla, AP during 2008–2010 and served as chairman, Board of Studies in Civil Engineering, Acharya Nagarjuna University, Nagarjuna Nagar, Guntur, AP, and also as member, BOS in Civil Engineering of Acharya Nagarjuna University.

Dr. Mohammad Kamrul Hasan completed a doctor of philosophy (PhD) in communication and network engineering from the faculty of Engineering, International Islamic University, Malaysia, in 2016. He was awarded and fully sponsored by the Malaysian International Scholarship (MIS) for his entire PhD program from the Ministry of Higher Education Malaysia. He completed his MSc in communication engineering studies in the area of communication and network engineering, International Islamic University, Malaysia. Currently Dr. Kamrul is working as a tenure-track assistant professor at the Center for Cyber Security, is faculty of information science and technology, the National University of Malaysia (UKM), and has world QoS ranking 141. In research, he has demonstrated excellent outcomes and received the gold medal award. For publication, he has been awarded the best paper award for many research papers. He has been published in 99 high impact factor journals recorded in Publon/Web of Science Researcher ID. Dr. Kamrul has been an experienced lead engineer for the supervision, installation, setup, and fault handling of PLC, WAMS, soft drive, controllers, MUDUBUS, and Merging Units. He has done fault handling and programming of the ladder logic design for the industrial LAN connected Programmable Logic Controller (PLC) devices. Bogra 50MW power plant is used for designing of Landis+Gyr 370. After 5 years of dedicated service, he resigned from the position in 2009 for further studies. Dr. Kamrul is a specialist well versed in elements on cutting-edge information-centric networks, smart grid communication systems and technologies, telecommunications, sensor systems and Industrial Internet of Things, ICT radio communication, data networking, and microcontroller-based electronic devices. Dr. Kamrul is a senior member of the Institute of Electrical and Electronics Engineers (SMIEEE-90852712), a Member of the Institution of Engineering and Technology (MIET-1100572830), and a member of the Internet Society (198312). Dr. Kamrul is a certified professional technologist (P.Tech.), Malaysia. He also served in the IEEE student branch as chair from 2014 to 2016. He has actively participated in many events/workshops/trainings for the IEEE and IEEE humanity programs in Malaysia. His application for professional engineer (CENG) is in progress.

Contributors

G. Anusha
Assistant Professor of
ECE Department
Sri Indu Institute of
Engineering and Technology
Hyderabad, India

T. Ashwini
Assistant Professor of
ECE Department
Sri Indu Institute of
Engineering and Technology
Hyderabad, India

Vinay Kumar Awaar
Department of Electrical and
Electronics Engineering
GRIET
Hyderabad, Telangana, India

P. Balamuralikrishna
Professor of ECE Department
CMR Technical Campus
Hyderabad, India

M. Bharathi
MB University
Tirupati, Andhra Pradesh, India

D. M. K. Chaitanya
Professor of ECE Department
Vasavi College of Engineering
Hyderabad, India

P. Srividya Devi
Department of Electrical and
Electronics Engineering
GRIET
Hyderabad, Telangana, India

M. Dharani
Sri Vidyanikethan Engineering College
Tiruapti, Andhrapradesh, India

Ajay Kumar Dharmireddy
Assistant Professor
Department of ECE
Sir C. Reddy College of Engineering
Eluru, India

Bandi Doss
Professor of ECE Department
CMR Technical Campus
Hyderabad, India

M. Greeshma
Assistant Professor
Department of ECE
Sir C. Reddy College of Engineering
Eluru, India

T. Venkata Krishnamoorthy
Sasi Institute of Technology & Engineering
Tadepalligudem, Andhrapradesh, India

B. Anil Kumar
Associate Professor
Department of Electronics and Communication Engineering
GRIET
Hyderabad, Telangana, India

Dayadi Lakshmaiah
Professor of ECE Department
Sri Indu Institute of Engineering and Technology
Hyderabad, India

C.H. Nagaraju
Professor of ECE Department
AITS
Rajampet, India

D. Prasad
Sasi Institute of Technology & Engineering
Tadepalligudem, Andhrapradesh, India

J. Praveen
Department of Electrical and Electronics Engineering
GRIET
Hyderabad, Telangana, India

V. Vijaya Rama Raju
Department of Electrical and Electronics Engineering
GRIET
Hyderabad, Telangana, India

G. Ramesh
Associate Professor
Department of Computer Science and Engineering
GRIET
Hyderabad, Telangana, India

L. Koteswara Rao
Professor of ECE Department and Principal
KL University Hyderabad
India

Ch. Madhava Rao
Assistant Professor
Department of ECE
Sir C. Reddy College of Engineering
Eluru, India

D. Nageshwar Rao
HOD & Professor of ECE Department
TKRCET
Hyderabad, India

R. Yadgiri Rao
Professor of H&S Department HOD
Sri Indu Institute of Engineering and Technology
Hyderabad, India

B. Ratnakanth
Assistant Professor of CSE Department
Anurag University
Hyderabad, India

K. Shashidhar
Professor
Department of ECE
Lords Institute of Engineering and Technology
Hydrabad, India

Annepureddy Sneha
Sri Indu Institute of Engineering and Technology
Hyderabad, India

Yerram Srinivas
Professor of ECE Department
Vignana Barathi Institute of Technology
Hyderabad, India

P. Srinivasulu
Assistant Professor
Department of ECE
Sir C. Reddy College of Engineering
Eluru, India

I. Venu
Assistant Professor of ECE Department
Sri Indu Institute of Engineering and Technology
Hyderabad, India

1

Improving IoT Security Using Blockchain

D.M.K. CHAITANYA[1], YERRAM SRINIVAS[2], B. RATNAKANTH[3], D. LAKSHMAIAH[4], AND G. ANUSHA[5]

[1]*Professor of ECE Department, Vasavi College of Engineering, Hyderabad, India*
[2]*Professor of ECE Department, Vignana Barathi Institute of Technology, Hyderabad, India*
[3]*Assistant Professor of CSE Department, Anurag University, Hyderabad, India*
[4]*Professor of ECE Department, Sri Indu Institute of Engineering and Technology, Hyderabad, India*
[5]*Assistant Professor of ECE Department, Sri Indu Institute of Engineering and Technology, Hyderabad, India*

Contents

DOI: 10.1201/9781003269991-1

1.1 Introduction

Internet of Things (IoT) technology means real things attached with the connection of inter-devices. It produces a huge quantity of information every day, in addition to information to be saved within the network. This network has to save information of restricted persons in addition to having a good message function and saving a huge quantity of information. As a result, people have problems protecting the information from hackers. Normally protection methods are very advanced; in this case, the user wants the latest technology.

In the IoT, each side is vulnerable to hackers. You can add an extra passage in addition to the crumple first side. The second one is the middle person, thwarting hackers in addition to the main face through secrecy, verification, and information reliability. When transferring data, the protection is contestable. Devices transfer the information to the services in a network, and the user receives information, which is very difficult for real devices.

Chain technology allows files and peripherals without the need for a central server system; in this, the middle person power force is deleted. Blocks have contacted one-to-one [7]; the next one is a spread ledger account. Everybody reviews the accounts to correct the records. The next one is the person who combines the blocks to chain. The user assigns the last one, or miner. It explains math values and assigns blocks to chain technology. It contains

Information: it is sent from the source to the destination
Before mess: each box has a value before it
Mess: each box has the natural mess value of information

1.1.1 Workproof

The latest box user Insert can evaluate math issues. The issue is too hard to explain. Suppose a burglar is capable of modifying the first block, but they have to modify each box for the reason that each box consists of a before mess.

1.1.2 Program

A program contains the latest boxes to insert, and the head is able to insert boxes. Normally the head is busy with selection in addition to users answering validation.

1.2 Regarding Work

D Lestari et al. [1–3] planned a power observing scheme that generates power expenditure information that is sent to the user. Power is vital in the direction of supply power energy utilized by people. People can decrease energy expenditure related to fast growth with the help of IoT.

BK Baraman et al. [4–7] planned a cost-efficient system for observing and calculating energy expenditure (Figure 1.1).

1.3 Connecting a Sharp Strength Device with Chain Structural Design

- The user takes the readings to sharpen the strength of the device.
- A block diagram related to the sharp strength device is exposed in Figure 1.2; information is present above the liquid crystal display.

Information is transferred toward sharp devices. With the use of strength devices, information is saved within the boxes and depends on the workproof. Workproof is a math system to crack the issue.

1.4 Sharp Strength Device

1.4.1 Components

- Arduino UNO

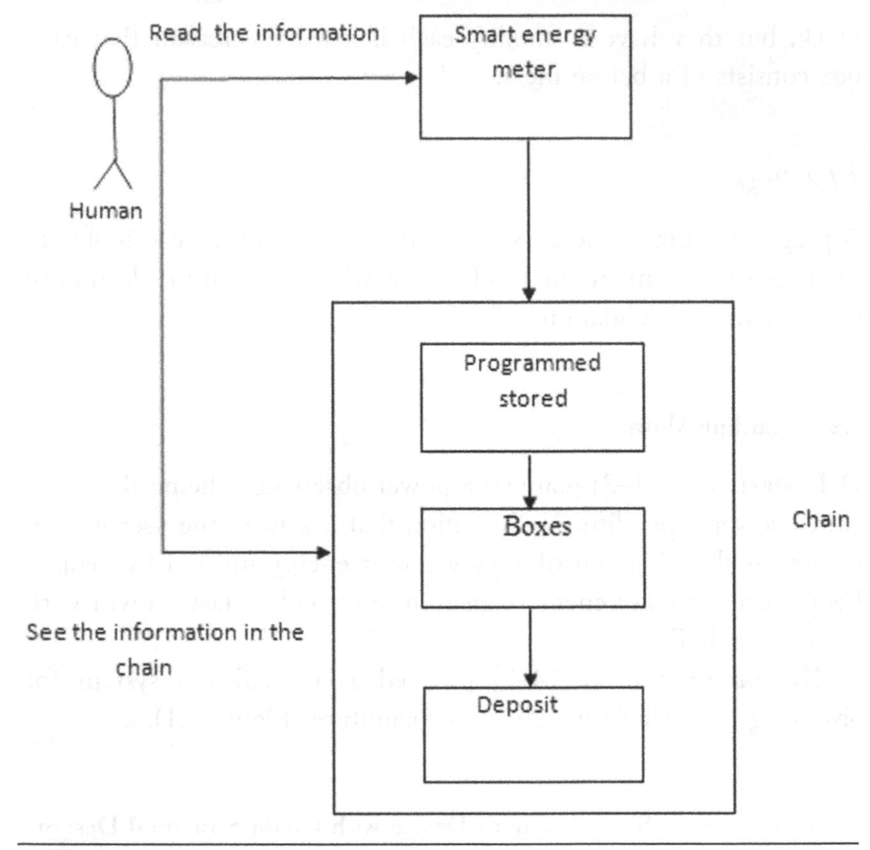

Figure 1.1 Sharp strength device contains boxes.

- LCD 16X2
- ACS712 Sensor
- 10k Ohm Potentiometer
- 220 Ohm Resistor
- Bread Board

1.4.2 ATmega328p Microcontroller

The Microcontroller board is based on the ATmega328p; the board is designed with analog and digital input/output pins associated with various IoT devices. The board is already programmed with a boot loader that allows code without the use of external software.

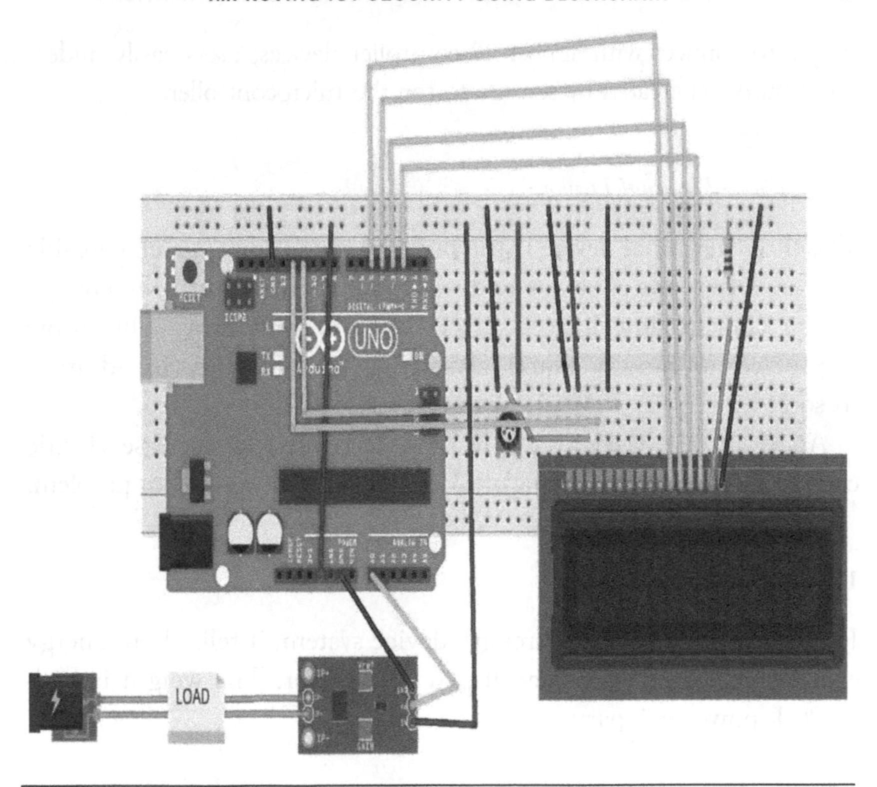

Figure 1.2 Sharp strength device schematic.

1.4.3 Liquid Crystal Display 16×2

The liquid crystal display 16×2 has two such lines and 16 characters on each line. The LCD has an output pin, cathode, anode, data configuration, one voltage pin, and one ground part, and the LCD 16×2 has memory devices: data memory and command memory. It has one read/write part and one enable part.

1.4.4 Sensing Device

The device has a low offset, linear hall circuit. The ACS712 current sensor starting voltage is a 5 V component speed and is designed from side to side, which is very near to a compelling indication toward the goal of the device, which is to convert into a physical quantity. The sensor finds the direct current and the altenative current. The ACS712 Sensor is simple to connect with almost all controller devices; users easily understand numbers created by sensors to see the microcontroller.

1.4.5 Three-Terminal Device

The three-terminal device is the potentiometer; one is a changeable end, and two are constant ends. A potentiometer utilizes change in the current flow instead of constant current flow. This topic is utilized to modify liquid crystal display intensity in addition to sound.

An electronic device is the resistor. It is used to decrease electric current and decrease 'V' forms of a diagram. It removes the kit problem.

1.5 Output

Figure 1.3 shows a sharp strength device system; it tells about energy expenditure of liquid crystal display "K" power. The weight is bulb 1,000 k power = 1 part.

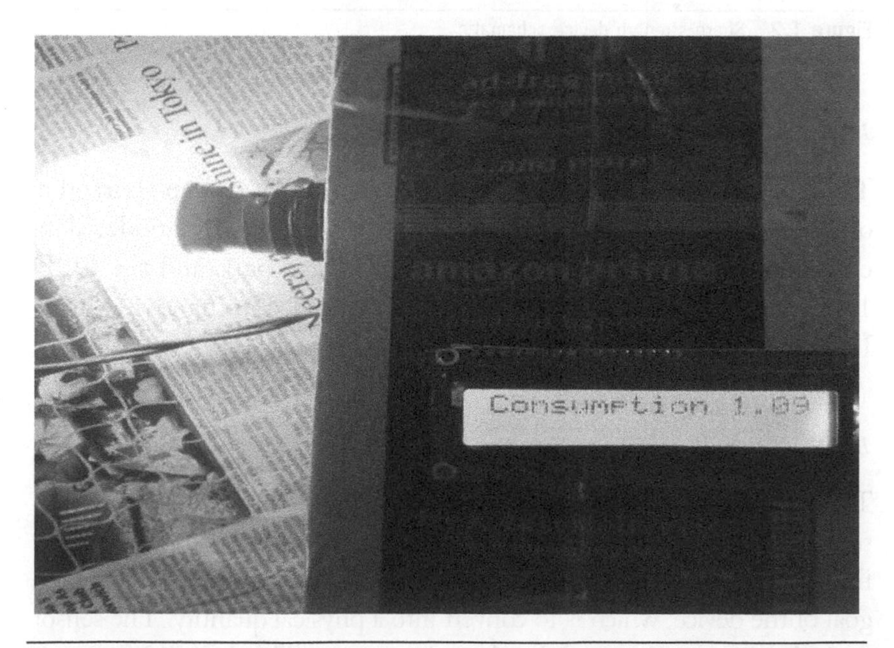

Figure 1.3 Sharp strength device.

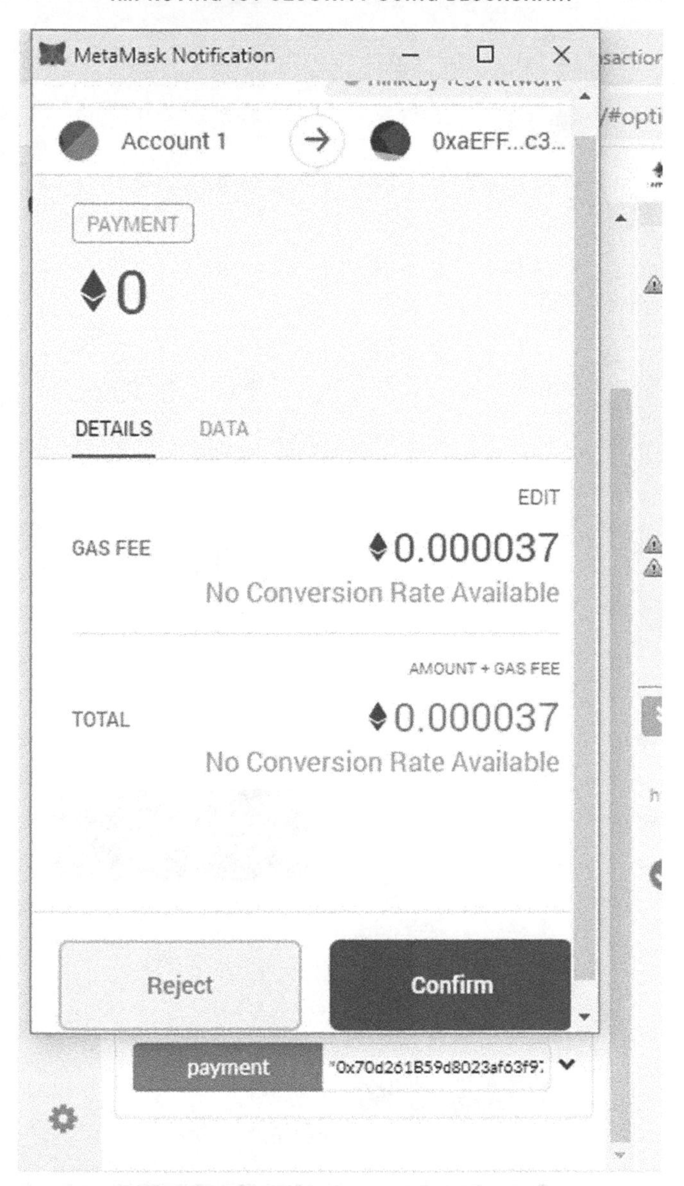

Figure 1.4 Sharp agreement.

The energy ability values within chain technology are in Figure 1.4. In Figure 1.4, each inquiry shows a small gas fee cost.

Figure 1.5 shows the customer's agreement; the above figure shows the quantity and decision of values within the agreement.

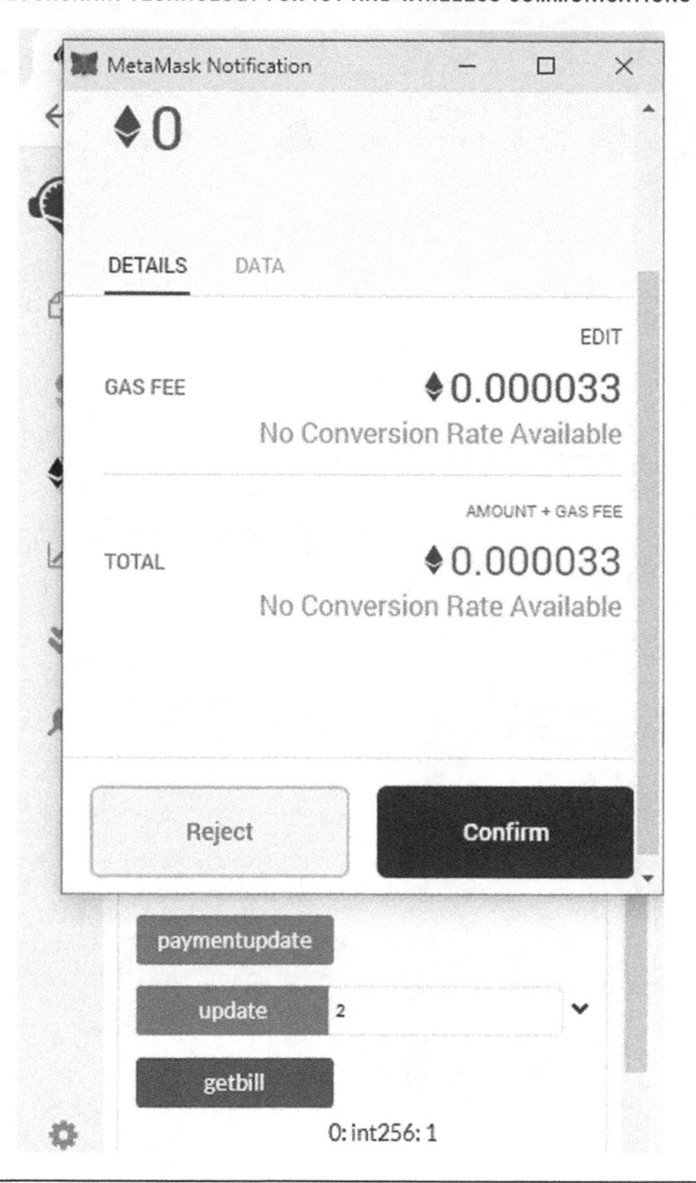

Figure 1.5 Compensation as of customer.

1.6 Conclusions and Further Work

Improving IoT security using blockchain can occur with a sharp strength device to give extra protection to people's information combined through chain knowledge. Power device force is skilled to gain power based on

TIME (HOURS)	READINGS (WATTHOURS)
0.1	2.78
0.05	0.11
0.2	9.1
0.15	5.99
0.3	14.95
0.25	12.13
0.4	20.69
0.35	17.78
0.5	25.75
0.45	23.19
1	30.84
0.55	28.29

Note: T1: Power expenditure values through weight

the weight. We introduced this topic of protecting information by means of chain technology to be produced effectively, as well as saving power versus the planned normal power.

References

1. Yaghmaee, M. H., & Hejazi, H. (2018, August). Design and Implementation of an Internet of Things Based Smart Energy Metering. In 2018 IEEE International Conference on Smart Energy Grid Engineering (SEGE) (pp. 191–194). IEEE.
2. Panarello, A., Tapas, N., Merlino, G., Longo, F., & Puliafito, A. (2018). Blockchain and IoT integration: A systematic survey. *Sensors*, 18(8), 2575.
3. Dorri, A., Kanhere, S. S., Jurdak, R., & Gauravaram, P. (2017, March). Blockchain for IoT Security and Privacy: The Case Study of a Smart Home. In 2017 IEEE international conference on pervasive computing and communications workshops (PerCom workshops) (pp. 618–623). IEEE.
4. Lestari, D., Wahyono, I. D., & Fadlika, I. (2017, November). IoT based Electrical Energy Consumption Monitoring System Prototype: Case Study in G4 Building Universitas Negeri Malang. In 2017 International Conference on Sustainable Information Engineering and Technology (SIET) (pp. 342–347). IEEE.
5. Barman, B. K., Yadav, S. N., Kumar, S., & Gope, S. (2018, June). IOT Based Smart Energy Meter for Efficient Energy Utilization in Smart Grid. In 2018 2nd International Conference on Power, Energy and Environment: Towards Smart Technology (ICEPE) (pp. 1–5). IEEE.

6. Kouicem, D. E., Bouabdallah, A., & Lakhlef, H. (2018). Internet of things security: A top-down survey. *Computer Networks*, 141, 199–221.
7. Mishra, J. K., Goyal, S., Tikkiwal, V. A., & Kumar, A. (2018, December). An IoT Based Smart Energy Management System. In 2018 4th International Conference on Computing Communication and Automation (ICCCA) (pp. 1–3). IEEE.

2

Blockchain-Based Secure Big Data Storage on the Cloud

BANDI DOSS[1], P. BALAMURALIKRISHNA[2], C.H. NAGARAJU[3], DAYADI LAKSHMAIAH[4], AND S. NARESH[5]

[1]*Professor of ECE Department, CMR Technical Campus, Hyderabad, India*
[2]*Dean R&D and Professor of EEE Department, Chalapathi Institute of Technology, Guntur, India*
[3]*Professor of ECE Department, AITS, Rajampet, India*
[4]*Professor of ECE Department, Sri Indu Groups, Hyderabad, India*
[5]*Assistant Professor of ECE Department, Sri Indu Institute of Engineering and Technology, Hyderabad, India*

Contents

DOI: 10.1201/9781003269991-2

2.1 Introduction

In the hasty expansion of sequence technology security, blockchain has the lead role at this time when it creates decentralized peer-to-peer structures. Blockchain linking is a distributed ledger [1] that traces all transaction information together, to avoid an intermediary. It also provides a massive sum of records inside one block and it also provides additional safety for data by using a hash technique so deleting records may not occur. Blocks are organized in a chain to frame blockchain knowledge where each block has both a secure hash and information about the previous block. Blocks are embedded with the previous hash [2] by a secure hash of the preceding block and it helps in identifying if the block is altered or not. Data inside a blockchain undergo finalization at some point the data are linked to the block, so data changes will not occur. Blocks are open access to the user connected to the blockchain network in which the user data is not open. Blockchain maintains consensus algorithms such as proof of work and proof of stake to store sensitive information. Big data and the present undergo some issues in safety, storage, sharing, and authenticating the data. Blockchain-based research challenges are identified to resolve the issues. This document has contributed to the analysis of the blockchain network in [3] big data storage and cloud safety.

2.2 Record of Blockchain Network

Electronic business transactions help to transfer the cash where it is easy to use devices like mobile and IoT devices while users are away

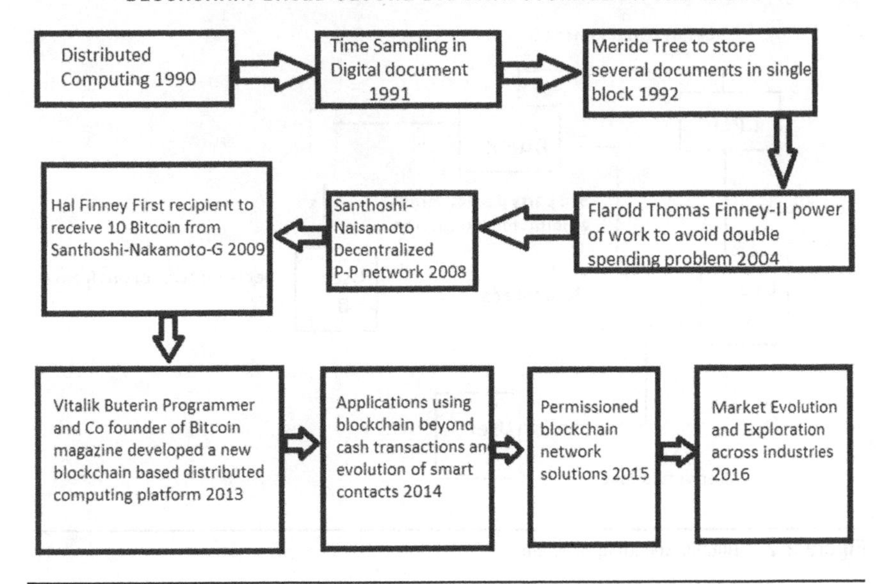

Figure 2.1 History of blockchain.

from the bank. Every electronic cash transfer is maintained secretly, even encrypting the transaction details of the user. Blockchain technology helps to ensure security for the B Pay for a fair contract, and no cash is detected from the user. The record of blockchain is depicted in Figure 2.1.

2.3 Requirement of Blockchain

2.3.1 High Processing Fee

The centralizer transmits between the dispatcher and recipient. Think about a transaction where a dispatcher transmits cash to a recipient where the arbiter charges an enormous fee to bear out a contract detail plus will store and plan through a third party.

2.3.2 Twofold Expenditure

Double spending is a major problem that arises with electronic cash transactions when a client sends the same electronic cash token to two beneficiaries with the intention of financial credit. Figure 2.2

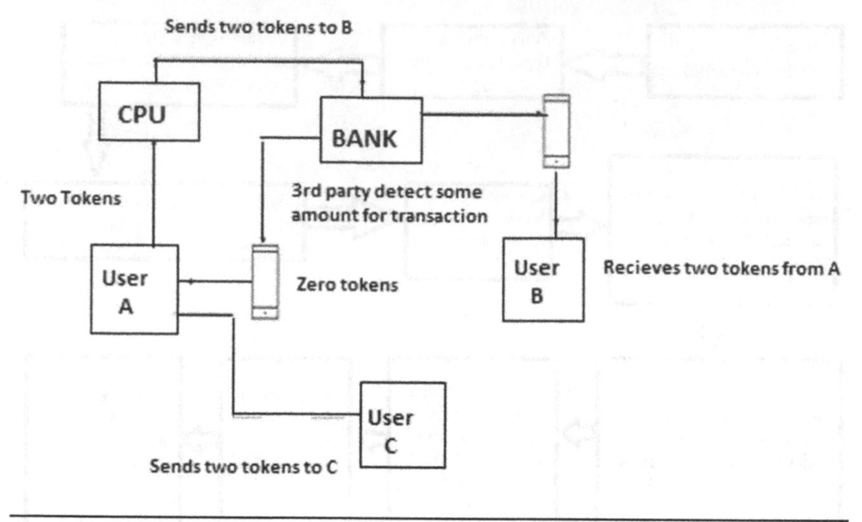

Figure 2.2 Double spending problem.

shows an electronic transaction wherein user A has two tokens. User A sends two tokens to user B (there is some problem in processing) so user A transfers for a second time the same two tokens to a user and double spending occurs.

2.3.3 Net Frauds

Unauthorized users are able to steal information about users through electronic transactions throughout net banking when third-party intermediaries use centralizers. Information lost is simply taken by hackers. In the Internet world, data manipulations and hacking are mainly done by hackers. Data changes plus alteration of information can happen during electronic transfer of data.

2.3.4 Fixing Corrupt Information

Data cleaning is a benefit designed for the user fix the corrupt data inside the centralized server; information transformation can take place repeatedly whilst reviewing data.

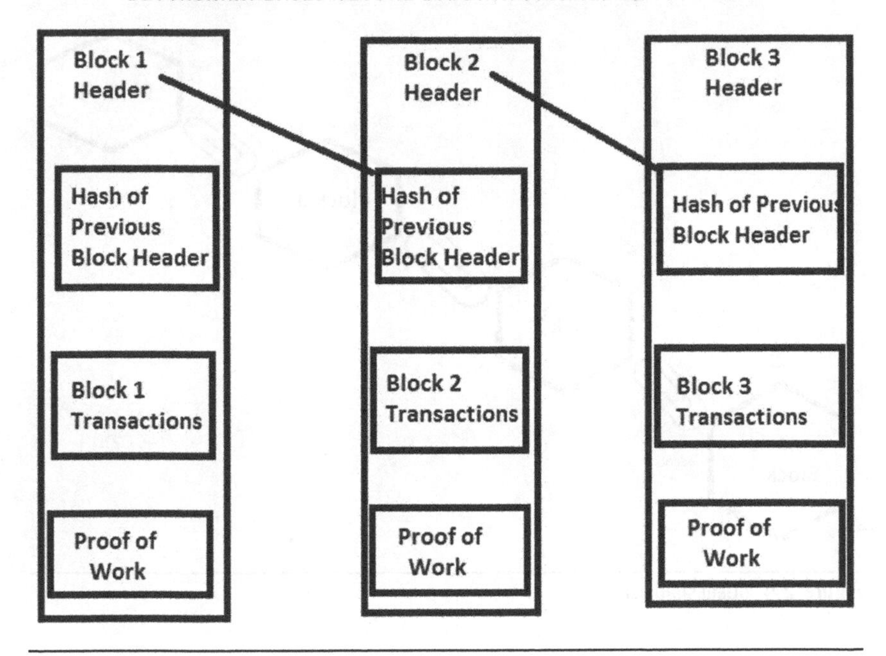

Figure 2.3 Overview of blocks in a blockchain.

2.4 Bitcoin vs Blockchain

Bitcoin is one of the data processing machine currency where it transfers and receives cash transverse centered peer-to-peer system. Processing fees in Bitcoin are small [4], compared to other third-party intermediaries such as banks and public sectors (Figure 2.3).

Every block in a blockchain has both its own hash and the previous block's hash where a hash consists of an alphanumeric value to identify the block in a blockchain. The primary chunk of the chain is named a genesis block, where the following block, does not have the hash, as is denoted in Figure 2.4.

2.5 Blockchain in Different Domains

A blockchain network is associated with the 'n' number of miners who mine the transaction.

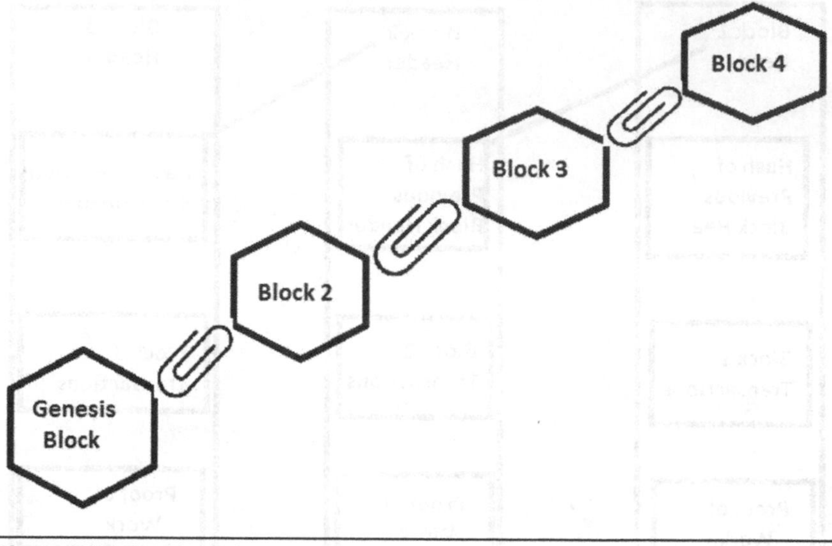

Figure 2.4 Genesis block.

2.5.1 Blockchain in Big Data

Big data deals with the huge amount of data collected, and it obeys the four V's, such as velocity, veracity, volume, and variety [5]. Data consumption of users increases day by day. The large volume of data increases, and a problem arises as this decreases the performance of the system because blockchain performs 'n' number of proceedings for block of in large information. The data transfer speed depends on time. When transferring a large amount of data, the transfer speed of transformation is going to be reduced to overcome this. We need to use blocks. Blocks will convert the big data into small blocks and improve the speed of data transfer. Speed has a major role to estimate the accuracy of information. Blockchain fights authentication trouble by verifying every block through the help of hashes where corrupt-data and the user are detached, while retrieving blocks. Blockchain stores incredible amounts of data in one single block, such as 'n' proceedings or a huge volume of information. Processing problems will occur while loading Big Data..

2.5.2 Blockchain Enumeration

Digital payment is a client and business-level common asset. They will access it every place [6]. Blockchain with protection solves the storage space copy in the chain and it provides improved security of information.

2.5.3 Blockchain in Internet of Things

Internet of Things is a machine that stores information gathered with an additional connection by means of cyber space through a sensor or wireless [7]. A combination of infrastructure among blockchain provides safety to the gathered data. It helps as long as there is security and time alone with the data.

2.5.4 Blockchain in Adaptable Device Enumeration

Adaptable device enumeration is the most well-liked automation in situations where people work together with another person. From first to last the machine and its application, let's say descriptions, will be stored in encrypted. Implementing mobile cloud using blockchain upgrades security of adaptable data.

2.6 Blockchain-Based Safety

Blockchain is a distributed ledger that will manage the absolute characteristic anywhere every block is linked to a structure in a blockchain system by a compromise algorithm. Pitman made the arrangement to data from blockchain on the condition that it provides security [8,9] by encrypting data.

It will provide storage space for the data as well. Blockchain technology also helps in distribution and authentication of the data by maintaining privacy.

2.7 Conclusion

This paper explored blockchain technology security for Big Data in the cloud. While analyzing Big Data, security and authorization are reviewed as well as the blockchain outcome. The blockchain system analyzed and addressed the research challenges to avoid data tampering. Blockchain provides security to the Big Data in the cloud for sharing and authentication of data, which resolves the storage space problem in the cloud.

References

1. Park, Jin Ho, and Park, Jong Hyuk (2017), 'Block chain safety in cloud enumerating: Use cases, challenges, and solutions', *Symmetry*, Vol. 9(8), pp. 164–177.
2. Turesson, H., Roatis, A., Las kowski, M., and Kim, H. (2019). 'Privacy-pre serving blockchain mining: Sybil-resistance by proof-of-useful-work', *ar Xiv pre print, ar Xiv:1907.08744.*
3. Dasgupta, Dipankar, Shrei, John M., and Gupta, Kishor Datta (2019), 'A survey of blockchain from safety perspective', *Journal of Bankingand Financial Technology*, Vol. 3(1), pp. 1–17.
4. https://bitcoin.org/bitcoin.pdf
5. Reyna, Ana, Martín, Cristian, Chen, Jaime, Soler, Enrique, and Manuel, Díaz (2018), 'On blockchain and its integration with IoT challenges and opportunities', *Future Generation Computer Systems*, Vol. 88, pp. 173–190.
6. Dorri, Ali, Kanhere, Salil S., Jurdak, Raja, and Gauravaram, Praveen (2019), 'LSB: A light weight scalable blockchainfor IoT safety and anonymity', *Journal of Parallel and Distributed Computing*, Vol. 134, pp. 180–197.
7. Restuccia, Francesco, D'Ore, Salvatore, Kanhere, Salil S., Melodia, Tommaso, and Das, Sajal K. (2019), 'Block chain for the Internet of Things: Present and future', *arXivpreprint,arXiv:1903.07448.*
8. Mangia, M., Marchioni, A., Pareschi, F., Rovatti, R., and Setti, G. (2019), 'Chained compressed sensing: A blockchain-inspired approach for low-cost safety in IoT sensing', *IEEE Internet of Things Journal*, Vol. 6(6), pp. 6465–6475.
9. Liu, Chunci, Xiao, Yinhao, Javangula, Vishesh, Hu, Qin, and Wang, Shengling (2018), 'Normal chain: A blockchain-based E-commerce', *IEEE Internet of Things Journal*, Vol. 6, (3), JUN2019.

3

Hyper Ledger Fabric Blockchain for Data Security in IoT Devices

DAYADI LAKSHMAIAH[1], L. KOTESWARA RAO[2], R. YADGIRI RAO[3], I. SATYA NARAYANA[4], AND ANNEPUREDDY SNEHA[5]

[1]*Professor of ECE Department, Sri Indu Institute of Engineering and Technology, Hyderabad, India*
[2]*Professor of ECE Department and Principal, KL University Hyderabad, India*
[3]*Professor of H&S Department, HOD, Sri Indu Institute of Engineering and Technology, Hyderabad, India*
[4]*Professor, Sri Indu Institute of Engineering and Technology, Hyderabad, India*
[5]*Assistant Professor of ECE Department, Sri Indu Institute of Engineering and Technology, Hyderabad, India*

Contents

DOI: 10.1201/9781003269991-3

3.1 Introduction

Strong security is necessary to keep records undamaged between Internet of Things (IoT) devices. There are many challenges in implementing data protection for IoT policy. The capability of an unauthorized user to use the desired system for attacks such as denial of service, and only the authorized users should be allowed to add information in a secure system with no holdup. It is very vital for the message to be classified to build convinced that facts cannot be altered or viewed elsewhere. In an IoT function such as tidy meter, one should focus to stay away from any bother due to impression foremost to severe loss. A result for data security, private in order violation and tampering of facts at the check supplier, after getting it. Blockchain originates as one of the rising skills to address these issues. The data can be disseminated transversely in the systems and the security of the transferred data can be achieved. Transactions are saved in the system as ledger records. Contract data is in blocks and connected cryptographically with strong hash encryptions. Each block stores previous data, as the present block comprises the hash. If a hacker tries to modify one block, then it is immune to do modifications. As the blockchain technology is distributed, if the data is crashed, the ledger is stuffed inside the other nodes. So, tampering and data loss are avoided.

Permission blockchains build a chain delimited by all standard, recognized foundations. The applicant contains an analogous core, but may not believe all extras fully. Authorization assists for protecting the commands among contestants. Authorization blockchain consists of consent protocols. These consensus protocols may be CFT or BFT. Conventional blockchain stays away from any intended corrupt codes. So every proceeding begins a function by making sure the ledger is verified. Interpretation and distribution through the interpret power meter and upload it to the server using a authorization blockchain network. A smart contract contains regulations for growing the dependability of users.

3.2 Related Work

Security is implemented at the design stage to avoid security concerns: threat taxonomy at different levels, a collection of safety and solitude provisions for web metering derivative support on the accessible threats, dealing with issues of records altered by middle attack, sophisticated metering with MICAz motes for statement between smart meters, the assault-free customer and malevolent customer to protect solitude in the communal system. Blockchain provides distinctiveness and security to the clients by parallel answer. It provides a protected link between two IoT devices using Ethereum blockchain platform. Two research studies were conducted on IoT devices with and without blockchain. We focus on concerns by the sub-model of IoT. Server failures, which are centralized, cause a single-point failure, and ethical hacking causes data tampering. A pub/sub architecture developed for blockchain conserves discretion of confidential data. Blockchain method uses a spread technique to provide protection and preserve the patient's health care proceedings, such as authentication, encryption, accessing steps to get the data in blockchain. An IoT server platform was used to address the vulnerabilities and pressure for security in Mysql's Mobius configuration. The information composed and broadcast strongly [1–12]. It dealt with finding of user personal data in the blockchain IoT environment to process the proof. The zero knowledge proof was developed, and ABAC on hyper ledger fabric blockchain framework for access control in IoT system was projected. A blockchain-based framework using Ethereum to maintain EMR was planned. The framework intended to conserve isolation of the patient data and right to use the medical records to the approved person.

3.3 Preliminaries

The specific members are connected through a channel for specific transactions by providing security and discretion. In earlier systems, we have the order-execute architecture.

3.3.1 Hyper Ledger Fabric Architecture [13–17]

Hyper ledger fabric client SDK provides the structured libraries for chain code applications. The elements are described below.

3.3.1.1 Peers A number of peer nodes are in a blockchain system. The ledger and smart contracts are hosted by peers; they are considered as fundamental elements of the blockchain network. The instances of ledger and chain code are hosted by peers. Any transaction generated by a smart contract is recorded immutably in a ledger. In a blockchain network, the shared process is encapsulated by a smart contract, and shared information is encapsulated by ledgers. If the blockchain resources have to be accessed by application and administration, then they should have an interaction with peers since the ledgers and chain code are hosted by peers. Due to these reasons, peers are considered to be basic construction blocks of a hyper ledger fabric blockchain network. Peers of organization are connected through a channel. A peer performs many roles such as an endorsing peer, committing peer, anchor peer or a leading peer.

The endorsing peers are involved in executing a smart contract during a transaction, and they return a signed response back to the client application. The committing peers are involved in validating the blocks of transactions that are orderly arranged and apply blocks to the local ledger copy. Since all peers store a copy of the ledger, all peers in the network can take the role of committing peer. An anchor peer will be the first peer in the channel that will be discovered by other organizations on the network. If an institute has many peer nodes, important peers engage in conversing with others.

3.3.1.2 Blockchain Ledger A blockchain ledger has a database and blockchain. The collection of states is stored to assist the developer to reduce the work by checking the whole contract log. Blockchain holds deals, enclosed as interlinked. Deals are stored in each block to specify the data. It confines all updates and deals are accrued inside and added to it.

The blockchain data cannot be customized. The data are changed when updates take place. A blockchain consists of a chain of block dealings that are unchallengeable.

3.3.1.3 Elegant Agreement Right of entry has many laws. If a customer requests data, it should be provided.

3.3.1.4 Orderer Nodes A local replica of the ledger is stored in blocks. An ordering service is a collection of ordered nodes within the network, and there will be a single ordering service for a network. The policies of the channel and membership information of each member of the channel are maintained in a channel configuration. The ordering service will have the channel configuration for the network, and hence they administer a network.

3.3.1.5 Network Policies The official document power offers the acquiescence proof for a business to validate the system. The customer request applies a credential to prove a business proposal to support a business suggestion and append a transaction to the ledger.

3.3.1.6 Channel A channel is the secure communication link between the members by creating a particular channel that can communicate, isolate data, and maintain confidentiality.

3.3.1.7 Identities and MSP X.509 digital certificates have identities, that are used to determine the particular actor permissions to access resources and information. MSP provides the policies that govern valid identities for organization. The X.509 certificates are used as identities in implementation of MSP in fabric. The MSP lists the identities to define the members of an organization.

3.4 System Model and Design

In IoT system architecture, blocks are blockchain network, web server, web client, mobile client, and Arduino client (smart energy meter) (Figure 3.1).

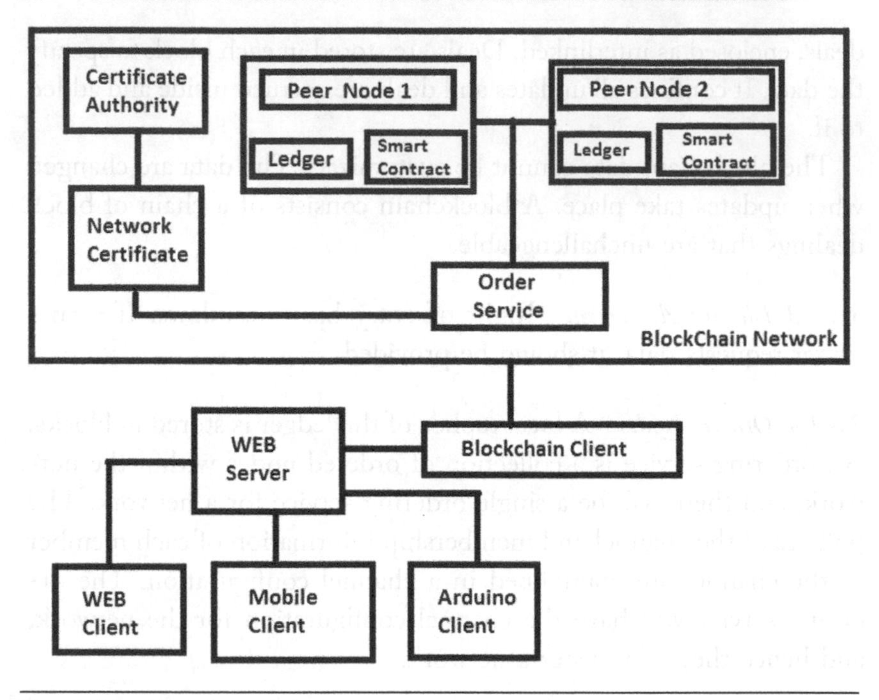

Figure 3.1 IoT system overview.

Implementation steps:

1. Create web server and host APIs.
2. Set up contact between the IoT sensor machine and the server.
3. Create a web client and manage admin activities.
4. Create a mobile client for registered users.
5. Set-up of connections with the web server.

3.4.1 Blockchain Network

Certificate authority (CA) issues the certificates for actors to authenticate to the network. The peers, orders etc are the active elements provides/use digital identities. X.509 certificate have the permissions and used in implementation and reorganization of MSP from a authorized source.

According to network policy, a network constructed should have a single order node and one peer for two organizations. The digital certificate is issued to the participants. The permission is granted to the linked channel. The network maintains two ledgers for user data and usage data. A single smart contract with multiple functions runs on peers.

3.4.1.1 Web Server It is a system program that serves web pages for users. The web server processes and provides a web page to the client. In azure, cloud system requirements will change as the size of the blockchain changes. The system requirements are two core CPU, 4GB memory, 10 GB of HDD/SSD, and Linux-based OS. The application is divided into UI routes and API routes. The API routes start with the path/API. POST, GET for a user ID data, are dependent on the blockchain module. The blockchain module is packaged as a Javascript module and is imported using RequireJS pattern. All are keen with essential Javascript constructs and exported as functions. The respective REST APIs are programmed to switch the queries and chant requests (Figure 3.2).

3.4.2 Web and Mobile Client

They fetch the information from the server and provide user interface. Mobile application have been developed. A mobile client can only fetch in sequence of an exacting user. The web client is provided with contact to analyze all users' information and also with the right to

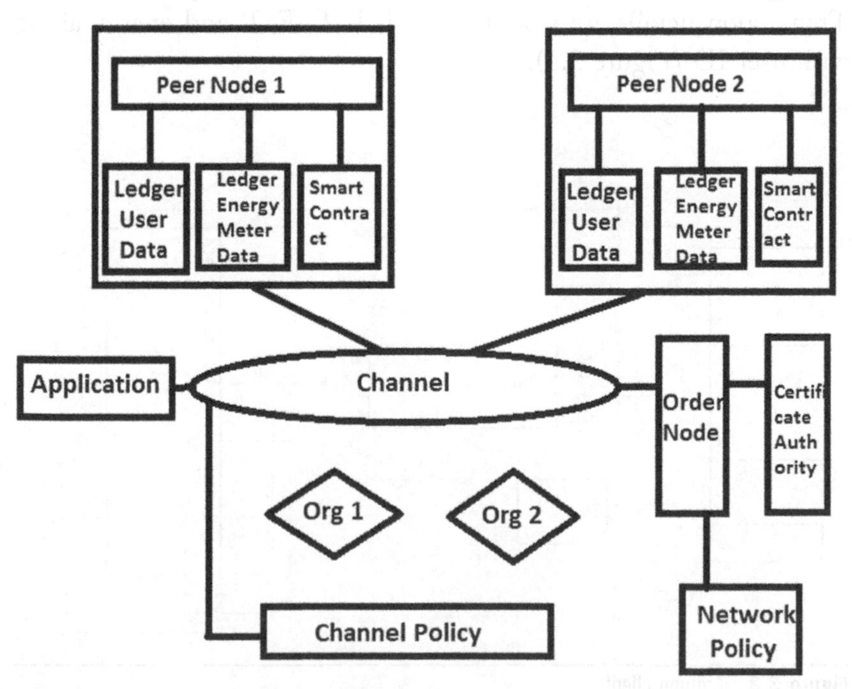

Figure 3.2 Blockchain network.

use for creation of new users. A user ID for every new user is created to generate transactions.

3.4.3 Arduino Client

The Node MCU acts as an Arduino client, which reads the energy meter data through a serial port and posts this data to the web server. SDM120M is used as the energy meter, which is capable of measuring the voltage in volts (V), current in amperes (A), power in watts (W), frequency in hertz (Hz), energy in KWh, power factor etc. of the connected load (Figure 3.3).

SDM120M is for reading the measured value. SDM120 with RS485 is used to communicate with systems using the Modbus RTU Protocol. It uses a MAX485 TTL – RS485 board and provides two-way serial communication signal conversion between the RS485 to TTL and vice versa.

3.5 Results and Analysis

Transaction details are stored with V, I, T, F, P and energy along with user ID (Figure 3.4).

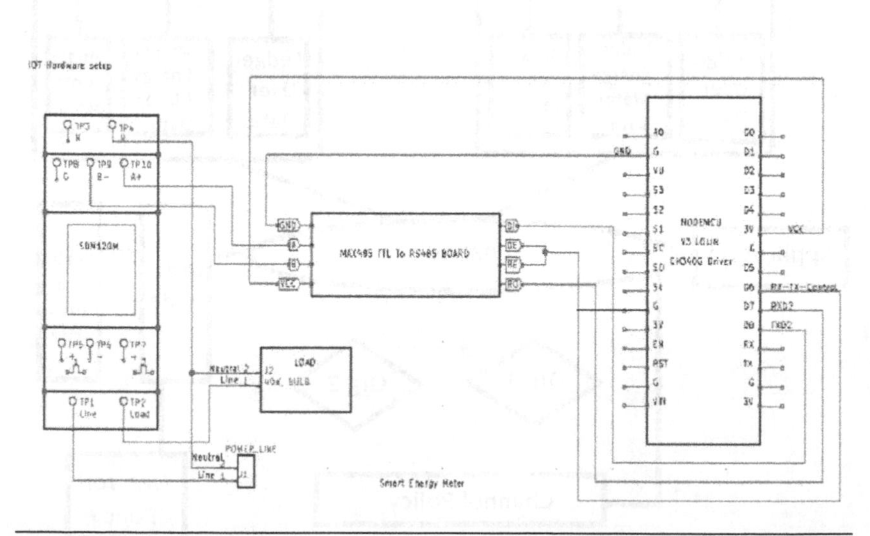

Figure 3.3 Arduino client.

```
{
  "_id": "USAGE00041587012828673",
  "_rev": "1-00c130f2aa57530ef23b232ea7a117ef",
  "current": "0.17",
  "docType": "usage",
  "energy": "0.36",
  "frequency": "50.09",
  "power": "39.80",
  "time": "1587012828644",
  "userId": "USER0004",
  "voltage": "231.40",
  "~version": "CgMBMQA="
}
```


Figure 3.4 Details of data in one of the transactions.


```
{
  "_id": "USER0000",
  "_rev": "3-bc4e13a8da8a857087fdbc03ec914bd9",
  "docType": "user",
  "userId": "USER0000",
  "userName": "Ronaldo",
  "~version": "CgMBBwA="
}
```


Figure 3.5 The detailed transaction record of USER0000 reflected in peer0 of Org1.

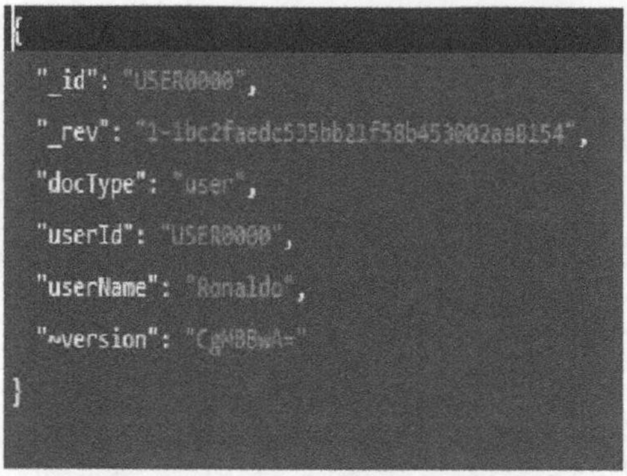

```
{
"_id": "USER0000",
"_rev": "1-1bc2faedc535bb21f58b453002aa8154",
"docType": "user",
"userId": "USER0000",
"userName": "Ronaldo",
"~version": "CgMBBwA="
}
```


Figure 3.6 The detailed transaction record of USER0000 reflected in peer0 of Org2.

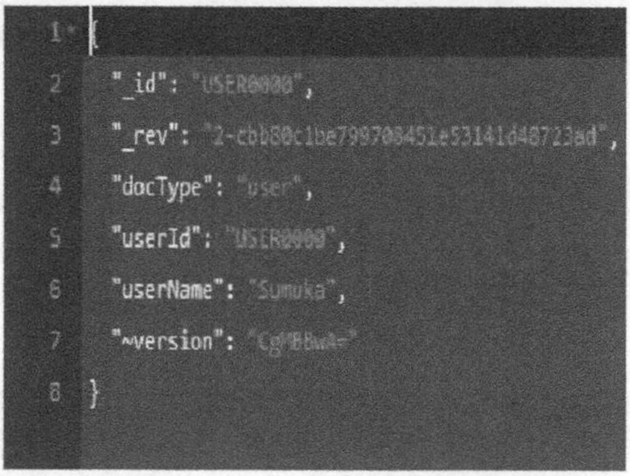

```
1 {
2   "_id": "USER0000",
3   "_rev": "2-cbb80c1be799708451e53141d48723ad",
4   "docType": "user",
5   "userId": "USER0000",
6   "userName": "Sumuka",
7   "~version": "CgMBBwA="
8 }
```


Figure 3.7 Transaction details in peer0 of Org1 after modifying username.

On observing the transaction records, both peers' data are the same. So they are decentralized and distributed (Figures 3.5 and 3.6).

To ensure the safety of data, anybody who tampers with the peer data, the original information in another peer, thus provides the protection of data (Figures 3.7 and 3.8).

The data are modified in the peer with username, and results are on the mobile app (Figures 3.9 and 3.10).

In blockchain, height increases after each write operation (Figures 3.11 and 3.12).

```
{
  "_id": "USER0000",
  "_rev": "1-1bc2faedc535bb21f58b453002aa8154",
  "docType": "user",
  "userId": "USER0000",
  "userName": "Ronaldo",
  "~version": "CgMBBwA="
}
```

Figure 3.8 Transaction details in peer0 of Org2 after modifying username in peer0 of Org1.

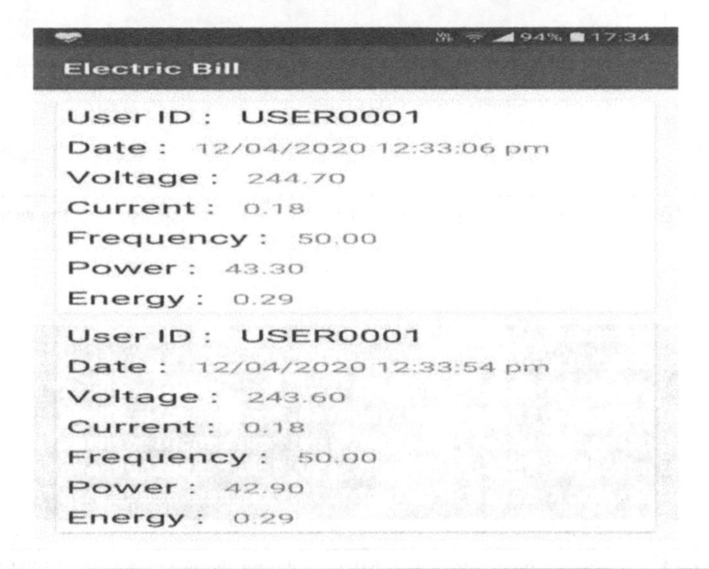

Figure 3.9 Mobile client sending a GET request and obtaining a response from web server.

3.5.1 Performance Analysis

The application was produced and examined for performance by POST and GET requests. The performance was examined on different platforms (Figures 3.13–3.15).

Multiple test tools were used for specific information to update the request. In a single threaded application, sequential tests executed the request, and the proposed testing measure average time for a transaction was identified as well as the actions for contemporaneous requests.

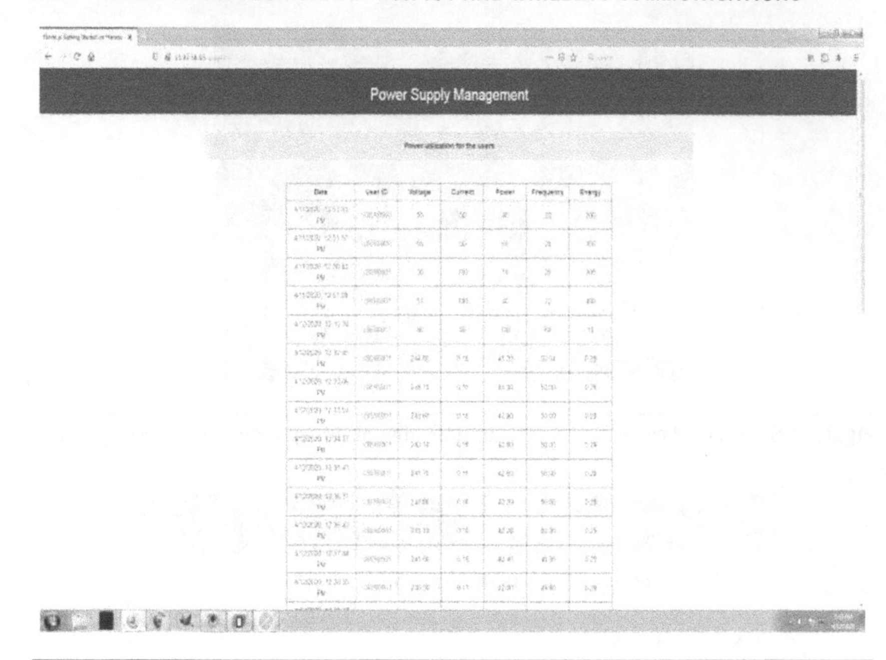

Figure 3.10 Web client sending a GET request and obtaining a response from the web server.

```
Blockchain info: {"height":65,"currentBlockHash":
Blockchain info: {"height":66,"currentBlockHash":
Blockchain info: {"height":67,"currentBlockHash":
Blockchain info: {"height":68,"currentBlockHash":
Blockchain info: {"height":69,"currentBlockHash":
Blockchain info: {"height":70,"currentBlockHash":
Blockchain info: {"height":71,"currentBlockHash":
```

Figure 3.11 Blockchain hash information before updating the usage details of USER0009.

```
Blockchain info: {"height":65,"currentBlockHash":'
Blockchain info: {"height":66,"currentBlockHash":'
Blockchain info: {"height":67,"currentBlockHash":'
Blockchain info: {"height":68,"currentBlockHash":'
Blockchain info: {"height":69,"currentBlockHash":'
Blockchain info: {"height":70,"currentBlockHash":'
Blockchain info: {"height":71,"currentBlockHash":'
Blockchain info: {"height":72,"currentBlockHash":'
Blockchain info: {"height":73,"currentBlockHash":'
```

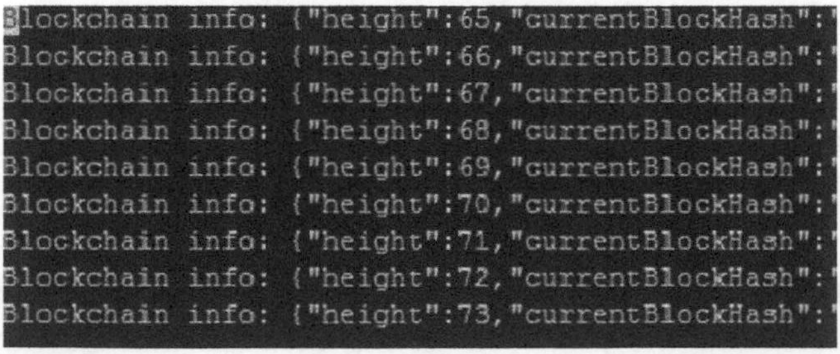

Figure 3.12 Blockchain hash information after updating the usage details of USER0009.

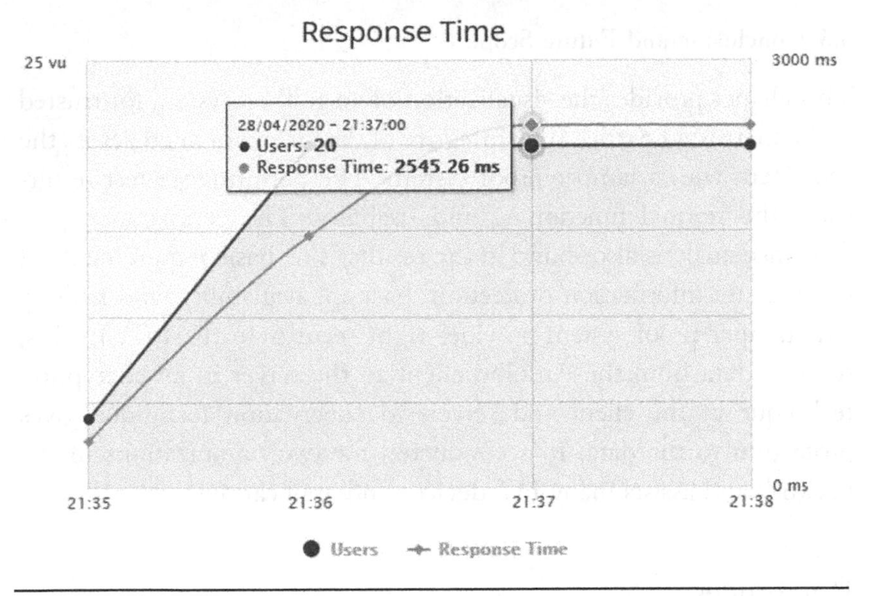

Figure 3.13 Sequential test results for 100 POST requests within 30 seconds using postman.

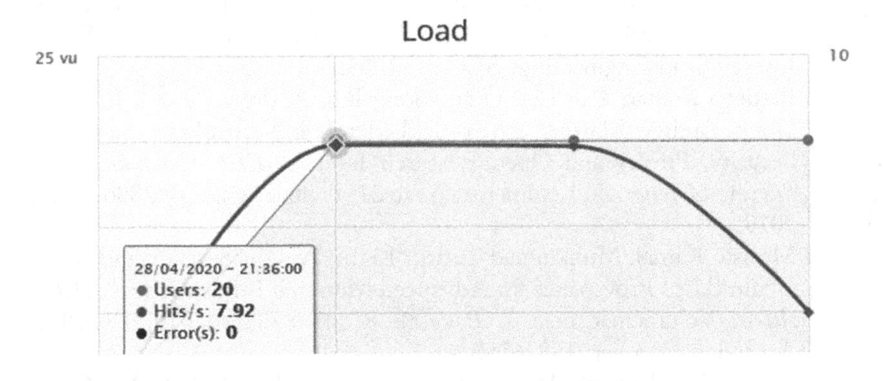

Figure 3.14 Load test graph for concurrent GET request using Blazeter tool.

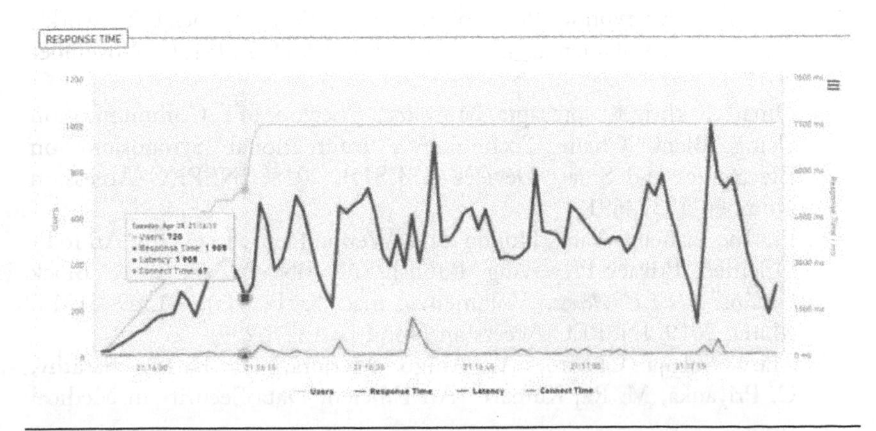

Figure 3.15 Response time graph for concurrent GET request using Blaze meter tool.

3.6 Conclusion and Future Scope

This chaper provides the visualization of an IoT ecosystem for trusted and non-trusted parties. The integrity of data is maintained across the ecosystem with a tamper-proof system. The performance test results show the normal functioning and usability. The comparative performance analysis also shows in the results. The basic requirements of the IoT are information protection, backup, availability, and scaling. The tamper-proof system provides tight security in the IoT. In this, sending data from the Arduino client to the server in an encryption technique at the client and server-side decryption technique gives protection to the data. It is conducted for two organizations in the network and assists many IoT devices and applications.

References

1. Obaid Ur-Rehman, Natasa Zivic, Christoph Ruland, " Security Issues in Smart Metering Systems", IEEE International Conference on Smart Energy Grid Engineering (SEGE), 2015.
2. Pardeep Kumar, Yun Lin, Guangdong Bai, Andrew Paverd, Jin Song Dong, Andrew Martin, "Smart Grid Metering Networks: A Survey on Security, Privacy and Open Research Issues", *IEEE Communications Surveys & Tutorials*, Volumemac_mac 21, Issue: 3, Page(s): 2886–2927, 2019.
3. Mohsin Kamal, Muhammad Tariq, "Light-Weight Security and Block Chain Based Provenance for Advanced Metering Infrastructure", *IEEE Access*, Volumemac_mac 7, Page(s): 87345–87356, 2019, INSPEC Accession Number: 18826750.
4. Ruiguo Yu, Jianrong Wang, Tianyi Xu, Jie Gao, Yongli An, Gong Zhang, Mei Yu, "Authentication with Block-Chain Algorithm and Text Encryption Protocol in Calculation of Social Network", *IEEE Access*, Volumemac_mac 5, Page(s): 24944–24951, 09 November 2017.
5. Dinan Fakhri, Kusprasapta Mutijarsa, "Secure IoT Communication Using Block Chain Technology", International Symposium on Electronics and Smart Devices (ISESD), 2018, INSPEC Accession Number: 18374691.
6. Pin Lv, Licheng Wang, Huijun Zhu, Wenbo Deng, Lize Gu, "An IoT-Oriented Privacy-Preserving Publish/Subscribe Model Over Block Chains", *IEEE Access*, Volumemac_mac 7, Page(s): 41309–41314, March 2019, INSPEC Accession Number: 18576298.
7. Mary Subaja Christo, A. Anigo Merjora, G. Partha Sarathy, C. Priyanka, M. Raj Kumari, "An Efficient Data Security in Medical

Report Using Block Chain Technology", International Conference on Communication and Signal Processing (ICCSP), 2019.

8. Jin HyeongJeon, Ki-Hyung Kim, Jai-Hoon Kim, "Block Chain Based Data Security Enhanced IoT Server Platform", International Conference on Information Networking (ICOIN), 2018, INSPEC Accession Number: 17720930.
9. Xi Peiyu, Zhang Qian, Wang Haining, Zhao Haoyue, Wang Chunyan, "Exploration of Block Chain Technology in Electric Power Transaction", International Conference on Power System Technology (POWERCON), 2018, INSPEC Accession Number: 18392665.
10. Chan Hyeok Lee, Ki-Hyung Kim, "Implementation of IoT System Using Block Chain with Authentication and Data Protection", International Conference on Information Networking (ICOIN), 2018, INSPEC Accession Number: 17720922.
11. Han Liu, Dezhi Han, Dun Li, "Fabric-IoT: A Block Chain-Based Access Control System in IoT", *IEEE Access*, Volumemac_mac 8, Page(s): 18207–18218, January 2020, Electronic ISSN: 2169-3536.
12. Eman-Yasser Daraghmi, Yousef-Awwad Daraghmi, Shyan-Ming Yuan, "MedChain: A Design of Block Chain-Based System for Medical Records Access and Permissions Management", *IEEE Access*, Volumemac_mac 7, Page(s): 164595–164613, November 2019, INSPEC Accession Number: 19144264.
13. https://hyperledger-fabric-ca.readthedocs.io/en/release-1.4/users-guide.html
14. https://hyperledger-fabric.readthedocs.io/en/release-2.0/key_concepts.html
15. https://hyperledger-fabric.readthedocs.io/en/release-2.0/build_network.html
16. https://kotlinlang.org/docs/reference/android-overview.html
17. Markus Schäffer, Monika di Angelo, Gernot Salzer, "Performance and scalability of Private Ethereum Block Chains", International Conference on Process Management, August 2019, Online ISBN 978-3-030-30429-4.

4

IoT-Based Concentrated Photovoltaic Solar System

V. VIJAYA RAMA RAJU AND PRAVEEN JUGGE

Department of Electrical and Electronics Engineering, Gokaraju Rangaraju Institute of Engineering and Technology, Hyderabad, India

Contents

4.1 Introduction

Solar cells are referred to as photovoltaic (PV) cells by scientists because they transform sunlight immediately into electrical energy. The PV effect is the procedure of transforming sunlight (photons) to electricity, and it is called voltage [1]. Bell Telephone scientists discovered the PV effect in 1954 when they realised that

DOI: 10.1201/9781003269991-4

Figure 4.1 A big silicon solar array on the roof of a business.

when silicon (a sand-like material) was subjected to sunlight, it created an electric charge. Solar cells were quickly being used to power spaceships and tiny electrical devices like calculators and watches (Figure 4.1).

Thousands of houses and businesses are powered by individual solar PV systems. PV technology is also being used in major power plants by utility corporations. Solar panels for homes and businesses typically consist of 40 solar cells grouped together in modules [2–3]. A typical home is powered by solar panels ranging from 10 to 20. The panels can look toward the south at a permanent angle or on a tracking system that follows the sun to catch as much light as possible. A solar array is a collection of solar panels that works together to create a complete system (Figure 4.2).

Hundreds of solar arrays are linked together to form a large utility-scale PV system for use by a big power utility or industry. Traditional solar cells are constructed of silicon, have flat plates, and are the most efficient [4,5]. Second-generation solar cells made of amorphous silicon or non-silicon materials like cadmium telluride are known as thin-film solar cells. Thin-film solar cells make use of semiconductor layers that are only a few micrometres thick. Because of its adaptability, thin-film solar cells can be utilised as roof shingles and tiles, building facades, or skylight glass (Figure 4.3).

Figure 4.2 Ground-mounted solar array in the fields.

Figure 4.3 A home's roof using thin-film solar tiles.

Solar inks, solar dyes, and conductive polymers are among the novel materials being utilised to manufacture third-generation solar cells utilising classic printing press technology [6,7]. Some modern solar cells utilise plastic lenses or mirrors to focus sunlight onto a very small piece of high-efficiency PV material. The PV material is more costly, but because these systems consume so little energy, utilities and industry are finding them to be more cost-effective. Concentrating collectors, on the other hand, are only utilised in the country's sunniest places since the lenses must be directed toward the sun.

4.1.1 Photovoltaic (PV) Technology Types

Solar radiation may be converted into electricity in two ways:

a. Photovoltaics (PV): Absorbed light is immediately transformed to electricity in some materials (photo effect).
b. Concentrated Solar Thermal (CST): To heat a liquid, direct light is focused into a single spot. The heat is then used to power a generator, much like in a traditional power plant (Figure 4.4).

4.1.2 Principle of Operation

The layers of a solar module are shown (Figure 4.5).

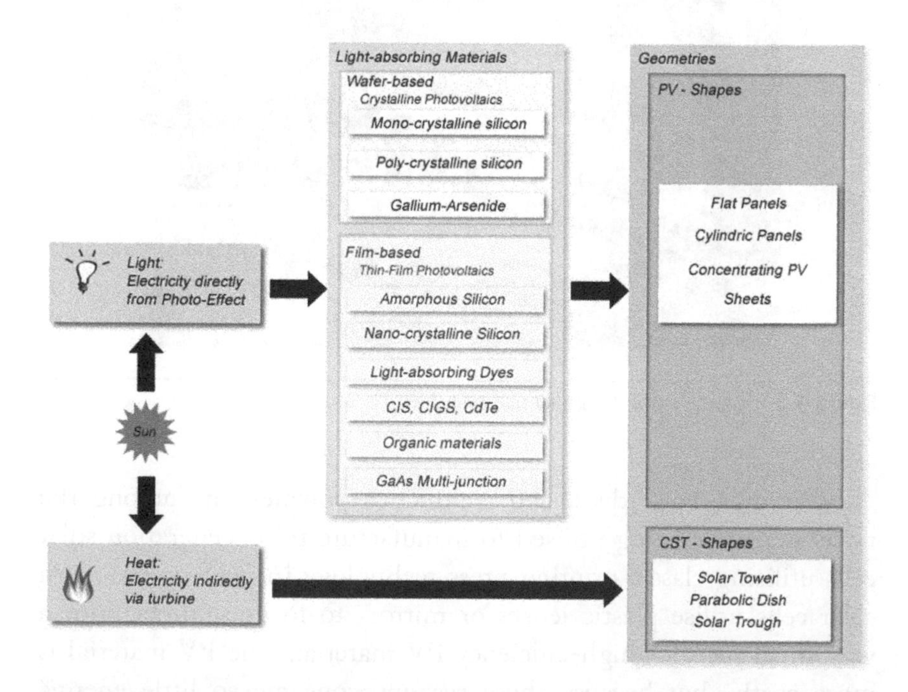

Figure 4.4 Photovoltaic (PV) technology types.

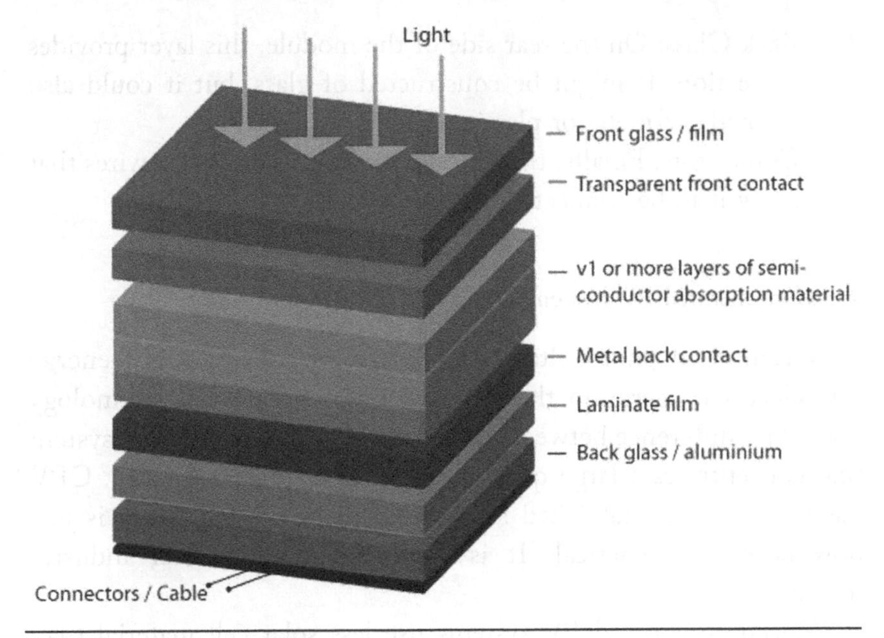

Figure 4.5 Solar module layers.

From the light-facing side to the back, all PV modules have a number of layers [8]:

a. Protection Layer: Glass is commonly used; however, transparent material can also be used in thin-film modules.
b. Front Contact: The front electric contact must be clear, or light will not be able to enter the cell.
c. Absorption Material: The layer where light is absorbed and transformed into electric current is the module's heart. All of the materials utilised in this project are semiconductors. In many cells, there is only one substance, which is usually silicon. However, many layers of various materials may be used to increase performance. Furthermore, all layers will be doped. In other words, each layer is subdivided into an n-doped and a p-doped zone. For additional information about doping, see the sections below.
d. Metal Back Contact: The electric circuitry is completed by a conductor at the rear.
e. Laminate Film: The construction is waterproof and heat-insulated thanks to the lamination.

f. Back Glass: On the rear side of the module, this layer provides protection. It might be constructed of glass, but it could also be made of metal or plastic.
g. Connectors: Finally, the module has connections and wires that allow it to be connected.

4.2 Concentrated PV Systems

A concentrated photovoltaic (CPV) system converts light energy into electrical energy in the same way as standard PV technology does. The difference between the two methods is the optical system that concentrates a large quantity of sunlight onto each cell. CPV has been a well-established science since the 1970s, but it is just now becoming practical. It is the most current solar industry invention.

Concentrator-based PV systems use less solar cell material than other PV systems. PV cells are the most costly component of a PV system per square metre. Using relatively affordable materials like plastic lenses and metal housings, a concentrator collects solar energy shining on a large area and concentrates it into a smaller region—the solar cell. One approach to evaluate the efficacy of this strategy is to look at the concentration ratio, or how much concentration the cell gets.

Compared to concentrator PV systems, flat-plate PV systems provide a number of benefits. For starters, concentrator systems reduce the number of cells needed, allowing for the use of more expensive semiconductor materials that would otherwise be prohibitively expensive in some designs. Second, when exposed to concentrated light, the efficiency of a solar cell improves. The design of the solar cell and the substance used to make it affect how much efficiency increases. Third, a concentrator can be made from tiny cells. Because large-area, high-efficiency solar cells are more difficult to manufacture than small-area cells, this is favourable.

Concentrators, on the other hand, face difficulties. To begin with, the focusing optics required are far more expensive than the simple covers required for flat-plate solar systems, and most concentrators require continuous monitoring of the sun throughout the day and year to be successful. As a result, reaching larger concentration ratios necessitates the use of both costly tracking methods and more precise controls. To concentrate light, both reflectors and lenses have been utilised.

The Fresnel lens, which focuses incoming light using a narrow saw tooth structure, is the most promising lens for PV applications. When the teeth are aligned in straight rows, the lenses serve as line-focusing concentrators. When the teeth are placed in concentric rings, light is focused at a central point. However, no lens can transmit 100% of the incoming light. Although lenses have a maximum light transmission of 90%–95%, most lenses transmit less. Additionally, concentrators are incapable of focusing diffuse sunlight, which makes up around 30% of the solar energy available on a clear day (Figure 4.6).

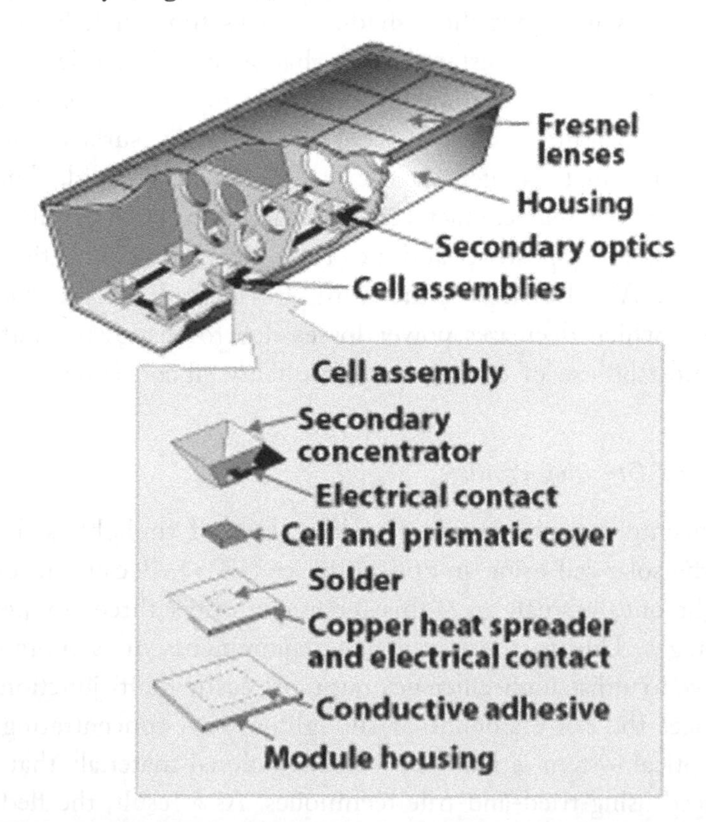

Figure 4.6 Arrangement of concentrators.

A typical concentrator unit consists of a lens that concentrates light, a cell assembly, a housing element, a secondary concentrator that reflects off-centre light rays onto the cell, a mechanism to dissipate excess heat created by concentrated sunlight, and various connections and adhesives. Heat is also a concern with high concentration ratios when photovoltaic (PV) systems utilise sun radiation. In addition, the quantity of heat produced is focused. As temperatures rise, cell efficiencies fall, and rising temperatures put solar cells' long-term durability at jeopardy. As a result, in a concentrator system, the solar cells must be kept cool, necessitating advanced heat sync cooling solutions.

One of the most significant design goals of concentrator systems is to minimise electrical resistance where the cell's electrical connections carry off the current generated by the cell. Fingers, or wide grid lines in the contacting grid on top of the cell, are helpful for low resistance, but their shadow blocks too much light from reaching the cell. Resistance and shadowing difficulties can be solved by prismatic coatings. These coverings work as a prism, directing incoming light to areas of the cell's surface that lie between the metal fingers of the electrical contact grid. Another alternative is a rear-contact cell, which differs from regular cells by having both positive and negative electrical connections on the back. All electrical connections are located on the back of the cell, which decreases power losses due to shadowing but also mandates the use of extremely high-quality silicon (Figure 4.7).

4.2.1 CPV Operating Principle

In concentrating photovoltaics, a large area of sunlight is focused onto the solar cell using an optical device (CPV). Because it focuses sunlight onto a small area, this technique offers three competitive advantages. Due to the lower area requirements, it is more cost-effective to utilise high-efficiency but more costly multi-junction cells to collect the same amount of sunlight as non-concentrating PV. The optical system is made up of conventional materials that were produced using tried-and-true techniques. As a result, the fledgling

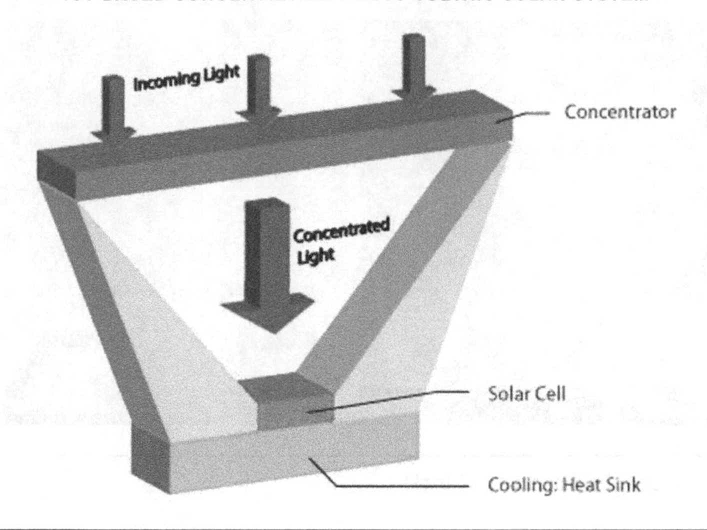

Figure 4.7 CPV operating principle.

silicon supply chain is less reliant on it. Additionally, optics are less costly than cells [9].

This strategy is confined to clear, bright situations because focusing light requires direct sunlight rather than diffuse light. It also means that in the great majority of circumstances, tracking is required. Despite the fact that it has been investigated since the 1970s, it is just now becoming a viable solar energy option. Because this is a novel technology, there is no single dominating design.

4.3 Construction and Operation of IoT-Based CPV

After the successful implementation of an incubated idea on a prototype, a CPV solar system with a dual axis solar tracking system was built for further enhancement of the efficiency. A concentrator can be rotated in Azimuth and elevation directions [10].

4.3.1 Mechanical Construction

This construction has mainly had three basic physical parts. They are

i. The base part and electrical connections
ii. The movable 'U'-shaped frame
iii. The concentrator and iron strip

Concentrated solar tracker (front and back).

Mechanical construction and electrical connections are described below.

4.3.1.1 The Base Part and Electrical Connections The base part is built in the form of square shape, and a rod is placed at the centre of the square parallel to its sides. A servo motor is placed just adjacent to the middle rod to make the gear interlocking for vertical axis rotation. Arduino Nano 33 IoT and the supply adaptor are placed on the edge side of the rod after soldering all the required connections, which is shown below.

The base part with the vertical axis rotation gear system with servo motor.

Gear system and Arduino Nano 33 IoT connections.

4.3.1.2 The Movable 'U'-Shaped Frame This is exactly a square-shaped 'U' frame to hold the parabolic dish with the help of screws on both sides. On one side, it has the servo motor with screws for horizontal rotation of the dish. This complete 'U' frame is placed on the middle rod of the base into one of the gears for movement.

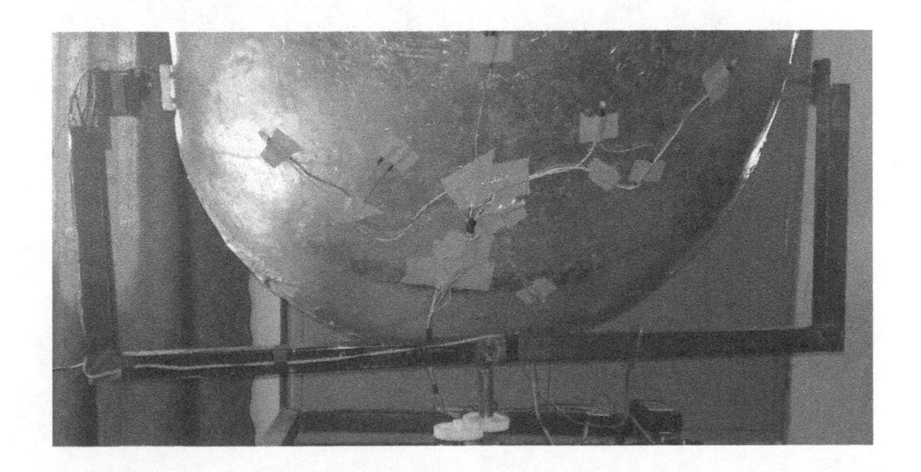

The movable 'U'-shaped frame mounted on the base.

Servo motor and mounted part on the gear.

4.3.1.3 The Concentrator and Iron Strip The concentrator has the eight light-dependent resistors placed on the diameters. Each perpendicular diametrical line on a parabolic dish has four sensors except at the centre. An iron strip is welded on both sides to the screws to support the PV panel. PV panel is placed on the inner side of the strip at the centre to absorb complete reflected light.

Concentrator and PV cell.

4.3.2 The Workings of IoT-Based CPV

- Initially, when the sunlight falls on the collector, the sun rays are reflected to the focused point where the solar panel is placed. If the panel is not placed at the proper place, it can be adjusted by the handle provided, along with the collector to hold the solar panel.

- When the sun rays are made to fall on the solar panel, the multi-junction solar panel has high efficiency compared to normal PV panels producing electricity.
- The LDRs placed on the collector for tracking the sunlight increase efficiency to a considerable value.
- Now the produced electricity is collected into a 12-volt battery, or it can be made to run a load since the voltage obtained is of DC from the multi-junction PV cells.
- There is an LCD display to show the amount of voltage produced by the multi-junction PV cells.
- We have a 12-volt, 5-watt solar module.

TECHNOLOGY	VOLTAGE (VOLTS)	CURRENT (AMP)
Concentrated solar power	20	0.326
Solar panel under sunlight	18.40	0.335
Solar panel under illumination	16.20	0.306

From the above experimental data, it is clearly seen that the voltage produced by the solar panel is more than the expected value when compared to the normal solar panel under sunlight.

4.4 Results and Discussion

The efficiency of the existing system is further improved by using dual-axis tracking and using the energy wasted in the form of heat in multi-junction PV cells for heating water. Instead of manual operation, tracking is done in a digital way by using Arduino Nano 33 IoT. Arduino Nano 33 IoT is used to collect the data from various light sensors attached to the panel and identifies the optimal location for irradiation, i.e., it tries to keep the collector at right angles to the sun rays. This will enhance the performance of the system. At the same time, the Nano will send the generation data to the cloud using the available Wi-Fi. This will enable the operator to monitor the performance of the entire system from a remote location.

4.4.1 Model Calculations Are Given Below

The solar panel ratings:

Max power output = 1 watt;
Max output voltage Vdc = 6 volts
I_{mp} = 167 milliamps.
Open circuit voltage = 7.2 volts;
Short circuit current = 183 milliamps
Power = Vmp*Imp = 6*.167 = 1 watt;
Energy = Power* Time
The typical voltage generated by panel when concentrated = 5.2 volts
The typical voltage generated by panel when diffused sunlight falls on panel = 4.5 volts

No. of units generated by the panel for five peak hours with focused sunlight (Vf)

= 5.2*.167*5 = 4.342 watt-hours/day.

No. of units generated by the panel for five peak hours with diffused sunlight (Vd)

= 4.5*.167*5 = 3.757 watt-hours/day.

Increase in efficiency = (Vf − Vd)/Vd × 100 = (4.342 − 3.757)/ 3.757 × 100
= 16.36%

It is clear that a concentrated PV solar system with a dual-axis tracking system is more efficient than the conventional fixed solar PV system (Table 4.1).

Table 4.1 CPV Solar System Performance Compared with Fixed-Tilt PV Solar System

DATE	CPV	FIXED TILT	INCREASED EFFICIENCY	DATE	CPV	FIXED TILT	INCREASED EFFICIENCY
01-11-21	3.87	3.26	15.85	16-11-21	6.1585	5.093	17.30129
02-11-21	3.30	2.86	13.40	17-11-21	4.6543	3.971	14.68105
03-11-21	4.62	3.85	16.70	18-11-21	2.6378	2.266	14.09508
04-11-21	3.52	2.97	15.64	19-11-21	1.4279	1.199	16.03053
05-11-21	3.80	3.22	15.28	20-11-21	3.2373	2.717	16.07204
06-11-21	3.54	3.00	15.23	21-11-21	5.1448	4.213	18.11149
07-11-21	3.60	2.99	16.82	22-11-21	5.2865	4.499	14.89643
08-11-21	3.22	2.75	14.48	23-11-21	6.1258	5.17	15.60286
09-11-21	4.15	3.62	12.86	24-11-21	4.1856	3.465	17.21617
10-11-21	4.58	3.72	18.79	25-11-21	3.9458	3.19	19.15454
11-11-21	3.52	3.04	13.77	26-11-21	4.6325	3.795	18.07879
12-11-21	4.13	3.54	14.26	27-11-21	3.5425	3.047	13.9873
13-11-21	3.76	3.19	15.17	28-11-21	1.7658	1.463	17.14803
14-11-21	5.26	4.29	18.51	29-11-21	0.9701	0.825	14.95722
15-11-21	6.10	5.09	16.56	30-11-21	2.8994	2.365	18.4314

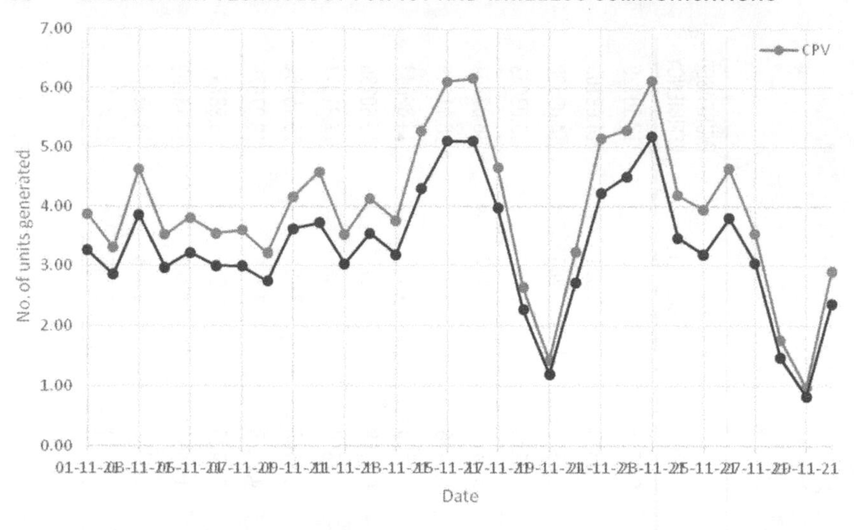

Daywise generation comparision between CPV sytem and fixed-tilt PV solar system.

4.5 Conclusions

Solar energy is abundant in nature and is available free of cost. It emits light and heat energy. The main objective is to use this energy to produce electricity. In this project, a concentrated solar system using mirrors, concentrators, tubes (for fluid flow), and multi-junction PV cells are used to improve the efficiency of the system. Heat energy is concentrated; this concentrated heat energy is used to produce power. On the other hand, multi-junction photovoltaic cells directly produce power using light energy of the sun. This project is to combine both systems and get more efficiency. This way, the thermal and light energy present in the sun rays are harnessed to generate more electrical energy. The addition of an IoT system with a dual-axis tracking system has further improved the system efficiency and provided an advantage to monitor the system from anywhere, anytime.

References

[1] George, M., Pandey, A. K., Abd Rahim, N., Saidur, R. "Recent studies in concentrated photovoltaic system (CPV): A review", 5th IET International Conference on Clean Energy and Technology (CEAT2018), 22–29.

[2] Tawa, H., Ota, Y., Inagaki, M., Mikami, R., Iwasaki, T., Ueyama, M., Nishioka, K. Comparison of CPV systems with lattice-matched and mismatched solar cells in long-term outdoor performance, 2018 IEEE 7th World Conference on Photovoltaic Energy Conversion (WCPEC), Waikoloa Village, HI, USA, 0961–0964. doi:10.1109/PVSC.2018.8547428

[3] Escarra, M. D. et al. A hybrid CPV/T system featuring transmissive, spectrum-splitting concentrator photovoltaics, IEEE 7th World Conference on Photovoltaic Energy Conversion (WCPEC), Waikoloa Village, HI, USA, 1658–1660. doi:10.1109/PVSC.2018.8547930

[4] Xu, Q., Ji, Y., Krut, D. D., Ermer, J. H., and Escarra, M. D. "Transmissive concentrator multijunction solar cells with over 47% in-band power conversion efficiency," *Appl. Phys. Lett.*, vol. 109, no. 19, 2016.

[5] Ji, Y. et al. "Optical design and validation of an infrared transmissive spectrum splitting concentrator photovoltaic module," *IEEE J. Photovoltaics*, vol. 7, no. 5, pp. 1469–1478, 2017.

[6] Vijaya Rama Raju, V., Mereddy, D. "Smart dual axes solar tracking system", 2015 IEEE International Conference on Energy Systems and Applications (ICESA 2015) organized by Dr. D. Y. Patil Institute of Engineering and Technology, Pune, India 30 October–1 November, 2015, sponsored by IEEE Pune Section.

[7] Elgeziry, M., Hatem, T. (2020). "Designing a dual-axis open-loop solar tracker for CPV applications". 2020 47th IEEE Photovoltaic Specialists Conference (PVSC), doi:10.1109/pvsc45281.2020.9300699

[8] Vossier, A., Chemisana, D., Flamant, G., Dollet, A. "Very high fluxes for concentrating photovoltaics – Considerations from simpleexperiments and modeling," *Renewable Energy*, vol. 38, pp. 31–40, 2012.

[9] Agrafiotis, C., Roeb, M., Konstandopoulos, A. G., Nalbandian, L., Zaspalis, V. T., Sattler, C., Stobbe, P., Steele, A. M. "Solar water splitting for hydrogen production with monolithic reactors," *Solar Energy*, vol. 79, no. 4, pp. 409–421, 2005. doi:10.1016/j.solener.2005.02.026.

[10] Bénard, C., Gobin, D., Gutierrez, M. "Experimental results of a latent-heat solar-roof, used for breeding chickens," *Solar Energy*, vol. 26, no. 4, pp. 347–359, 1981. doi:10.1016/0038-092X(81)90181-X.

5

Design and Implementation of an IoT-Integrated Smart Home System with End-to-End Security Using Blockchain Technology

G. RAMESH[1], B. ANIL KUMAR[2], AND J. PRAVEEN[3]

[1]*Associate Professor, Department of Computer Science and Engineering, GRIET, Hyderabad, Telangana, India*

[2]*Associate Professor, Department of Electronics and Communication Engineering, GRIET, Hyderabad, Telangana, India*

[3]*Professor, Department of Electrical and Electronics Engineering, GRIET, Hyderabad, Telangana, India*

Contents

5.1 Introduction

The concept of smart homes is gaining popularity in cities, as it provides smart access to the infrastructure of homes and it can be

DOI: 10.1201/9781003269991-5

controlled from a smart hand-held device from a remote place. Smart homes can help the owners and residents of a house to have accessibility to their homes and can control bulbs, doors, and other electronic devices from a distant place. It is possible to have smart cities as well due to the emergence of technologies like cloud computing, Internet of Things, and distributed technologies. The problem with existing smart home systems, as explored in [1–5], is that end-to-end security may be at stake in certain cases where security loopholes are exploited by adversaries. When security is compromised, it can cause many issues, as unauthorized people can enter into the home and do unexpected things. Therefore, there is a need for more security in smart homes.

IoT-integrated smart applications like smart homes create security and privacy challenges. They are as follows. Scalability is the main problem, as the current centralized IoT platforms have message-routing mechanisms that create bottlenecks in scaling up to a large number of devices used in IoT. There is the security problem that a huge number of devices are participating to generate data and such a setup may be subjected to distributed denial of service (DDoS) attacks. Lack of data standards is another cause of concern as it leads to challenges and interoperability problems. As IoT-integrated solutions are associated with a huge number of devices, cost is another important concern. The integration with a centralized cloud may prove to be a bottleneck in the case of any disruption of services from the cloud for any reason.

This is the rationale behind taking up this work, which is aimed at designing and implementing a smart home with IoT and cloud integration, besides the use of blockchain technology that provides end-to-end security. Blockchain, as explored in [6], is a technology that refers to a distributed ledger of transactions and peer-to-peer communication among participating nodes that meet security needs and address security challenges thrown by IoT-integrated smart applications. In the blockchain network, each participant is granted access to an up-to-date copy of the encrypted ledger so as to help the node to have read/write and validate transactions. Though blockchain is initially used in the financial domain, it is now gaining popularity and acceptance for end-to-end security in IoT-integrated smart applications. The vision of decentralized IoT is realized with blockchain technology.

It facilitates end-to-end secure transactions among the participating devices and coordination among them.

In this context, IoT and blockchain technology offer a promising solution to a smart home system, as the system can provide end-to-end security and overcome the problems aforementioned. The usage of an open-standard distributed IoT solution can solve many problems that are associated with centralized approaches. As the blockchain technology is nothing but a distributed ledger of transactions, it offers direct communication to connected devices. Such devices collect data, and that can be accessed by all legitimate participants. Thus, decentralized blockchain networks can provide improved security of IoT-based solutions. Blockchain technology ensures end-to-end security by executing predefined smart contracts and taking care of specific consensus mechanisms that identify actions of compromised devices. In essence, the blockchain-enabled IoT-integrated smart home system can secure devices and data collected by them. It is possible as all facility management suppliers participate in a private blockchain in a distributed environment to provide timely service and automate the activities related to security.

5.2 Related Work

This section provides review of literature on the related topics of the proposed work. In the literature, it is found that the smart homes concept has been around for some years. However, there lacks an end-to-end security guarantee due to the number of connected devices and diversified technologies that form IoT and lack of standardization. Smart home literature found in [1,2], [6–10] reveals this fact. Since the smart home concept is realized based on IoT technology, it is important to ensure that there is end-to-end security. The IoT integration with smart homes and other use cases is found in [3], [11–15]. It is also true that blockchain in the technology is in the distributed environment, and it can be easily integrated with IoT, as studied in [4], [16–20]. The blockchain technology that provides a distributed ledger of transactions is suitable for helping IoT devices to achieve security benefits, as discussed in [5,21], and [22]. The integration of an IoT smart home with blockchain provides an end-to-end security guarantee, and the smart home product that serves this privacy and security purpose is a need of society.

As explored in [1,2], [6–10], the smart home concept is not yet integrated with the sophisticated security infrastructure of blockchain technology. This is the reason, in the area of smart homes and smart IoT applications, of the need for an end-to-end security guarantee that can be provided by blockchain as per its claims. It is therefore essential to investigate the need for blockchain technology integration with IoT-enabled smart homes. The most relevant references found in the literature on the security issues of smart homes with present schemes are [1,2], [6–10]. They reveal the fact that security to smart homes is very important as loopholes in security can cause many issues. The emergence of blockchain technology provides a distributed ledger of transactions that is accessible to IoT devices or connected devices that participate in smart home IoT applications. There is a need for integration of blockchain technology with IoT-enabled connected devices of smart homes to ensure that both devices and the data collected by them are secured against privacy and security attacks.

5.3 Security Challenges

Smart homes with IoT and cloud integration have many security challenges and scalability issues. The reason behind security issues is that the environment is distributed in nature and thousands of devices may participate in the network. When any device is compromised or when security credentials are stolen, the whole system will be exposed to security risks. The existing solutions to the problem of security in smart homes are not scalable, and they do have loopholes like lack of standards and are prone to DDoS attacks and other attacks. They are not able to provide end-to-end solutions to the transactions in IoT-enabled smart homes. There is a need for end-to-end security in such systems. The aim of this chapter is to design and implement an IoT-enabled smart home system with blockchain technology for end-to-end security, irrespective of the make and platform of applications and devices that participate in distributed computing. Since blockchain is the distributed ledger of transactions that is accessible to all legitimate devices, it can help devices to be smart enough to prevent any security attacks. In other words, the smart home, with all its participating devices and data, is protected with blockchain technology integration.

5.4 Methodology

The methodology used in this chapter is described here. The research starts with review of literature to know the insights required to design and implement the proposed work (Figure 5.1).

In the analysis and design phase, the researcher will identify more accurate requirements and finalize them. Then the researcher designs the system. Afterwards, the design is converted into a working solution with a prototype application. Once a smart home prototype is built with IoT and cloud integration, it will be integrated with blockchian technology. After integrating with blockchain, the system is evaluated for end-to-end security and intended communications. The evaluation mainly focuses on device security and data security. Once the effectiveness of the blockchain with respect to end-to-end security is proved, the prototype is converted into an out-of-the-box commercial solution.

As shown in Figure 5.2, it is evident that the smart home devices and gateway of smart home are integrated with IoT and cloud platform using MQTT protocol. In turn, it is integrated with

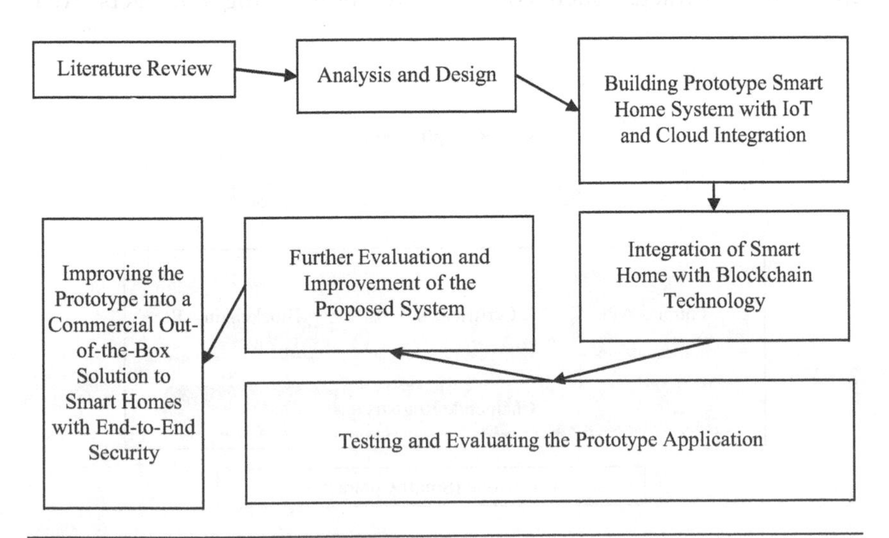

Figure 5.1 Conceptual design of the proposed research.

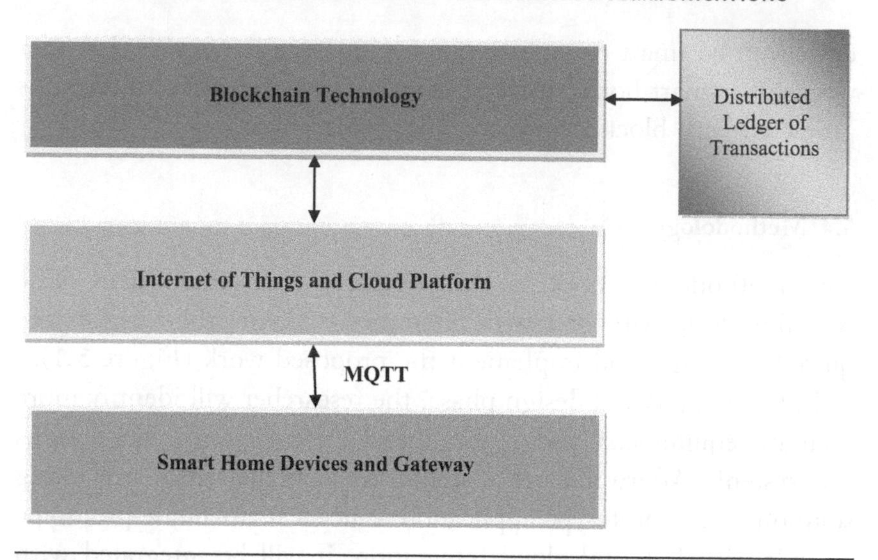

Figure 5.2 Outline of the proposed methodology.

blockchain technology using REST technology, which is interoperable in nature. The blockchain technology is equipped with a distributed ledger of transactions that is the important means of achieving end-to-end technology.

As shown in Figure 5.3, the blockchain technology has many components like Contracts API, Certificates API, Blockchain API, and Chaincode Registry. The security business logic is encapsulated in smart contracts. Such contracts are built using contracts API.

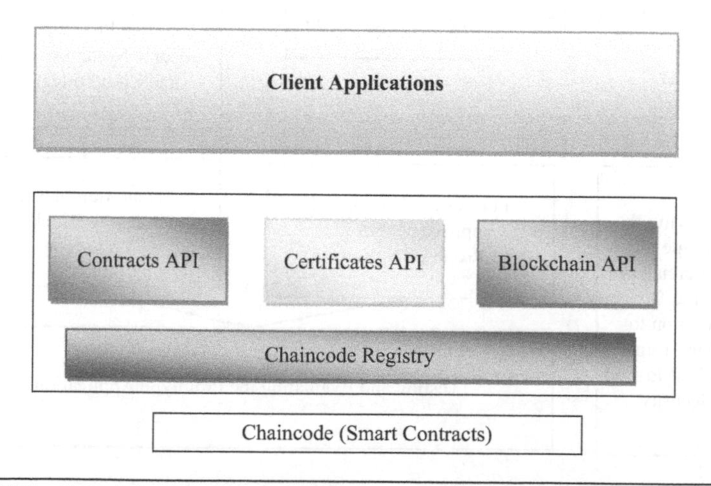

Figure 5.3 Blockchain technology components.

Blockchain API is the client API invoked by blockchain applications. Certificates API are used in the security process.

As shown in Figure 5.4, it is evident that the proposed system has integration with IoT, cloud, and blockchain technology. The smart home has plenty of electronic devices with sensors that are integrated with the IoT platform. The data collected by devices are sent to cloud storage. Such data are subjected to big data analytics to have essential business intelligence related to the usage of various resources at home and the behaviour of housemates. This business intelligence can help in making well-informed decisions. The devices participating in the computing are integrated with blockchain technology that enables storage of transactions in a distributed ledger. This ledger is made available to all participating devices so as to let them quickly validate transactions and identify compromised nodes, if any. The end-to-end security in the smart home is thus made possible.

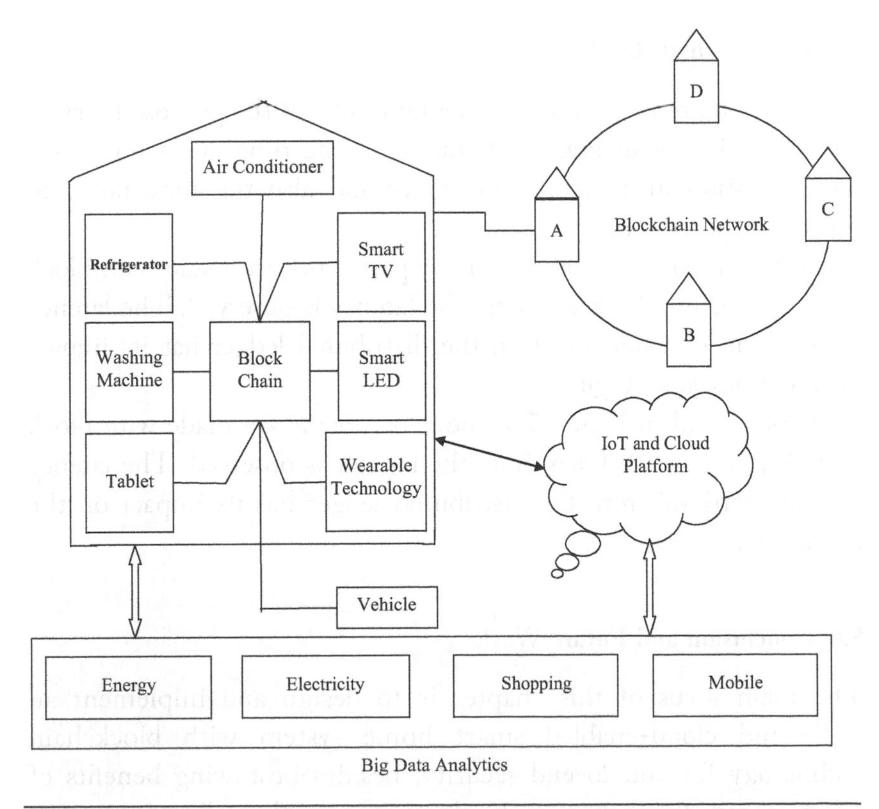

Figure 5.4 The proposed system with smart home system integrated with IoT, cloud, and blockchain technology.

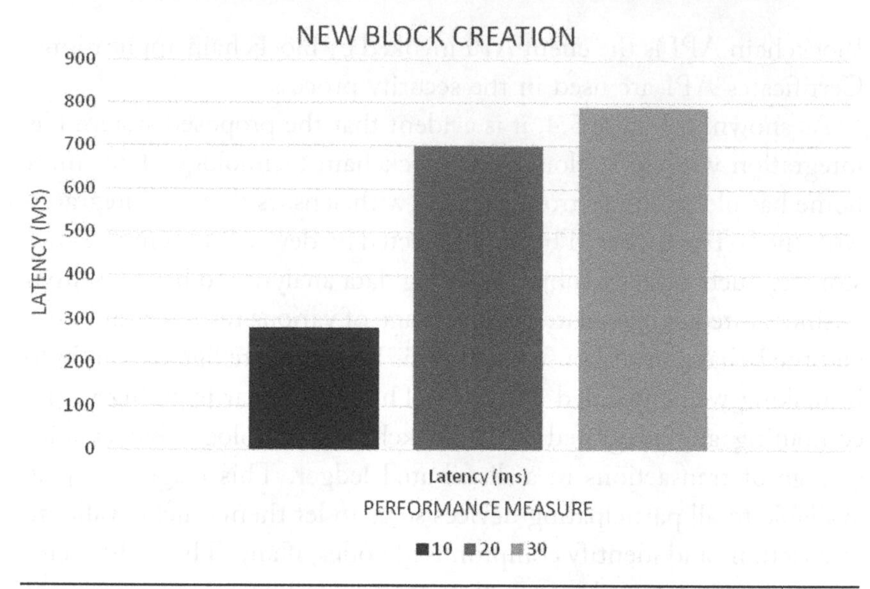

Figure 5.5 Shows performance of the system to create a new block in terms of latency.

5.5 Experimental Results

This section presents the experimental results of the proposed system with partial implementation. It measures the time taken for a new block creation in a distributed ledger and also the time taken to retrieve required data.

As presented in Figure 5.5, the experiments are made with block sizes 10, 20, and 30. Each time, the latency is observed. The latency for creation of a new block in the distributed ledger has its impact on the block size (Figure 5.6).

As presented in Figure 5.5, the experiments are made with block sizes 10, 20 and 30. Each time, the latency is observed. The latency for data retrieval from the distributed ledger has its impact on the block size.

5.6 Conclusion and Future Work

The main focus of this chapter is to design and implement an IoT- and cloud-enabled smart home system with blockchain technology for end-to-end security, besides bestowing benefits of smart home. Blockchain technology ensures that all participating devices in the distributed system can gain access to distributed

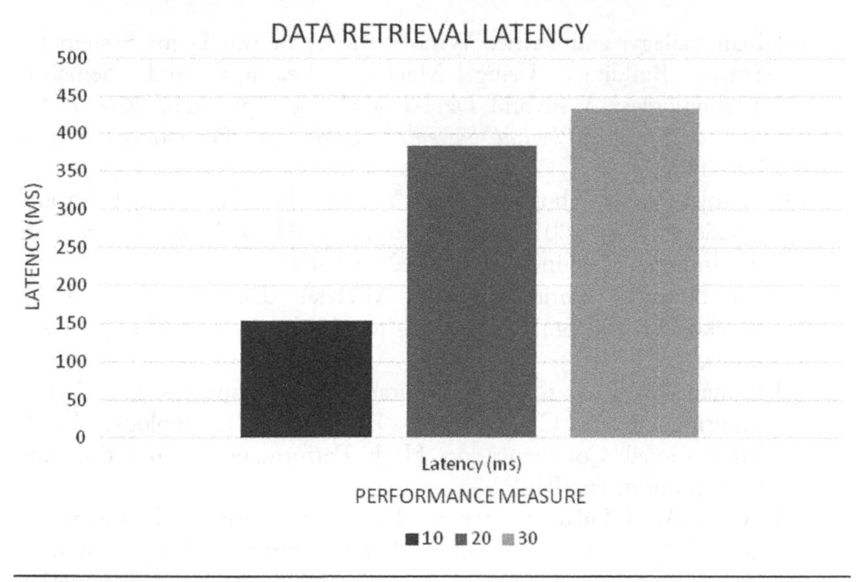

Figure 5.6 Shows performance of the system to retrieve required data.

ledger of transactions and quickly update and validate the transactions. In the process, they can easily detect compromised nodes. In this chapter, we proposed a methodology for smart home and blockchain integration for a higher level of security. The IoT-based smart home–related transactions are saved to a cloud-based distributed ledger in blockchain. The system has a provision to create a hash and encrypt the transactions prior to sending to the blockchain. An empirical study is made to have partial realization of the proposed system. The results revealed that the transactions in a smart home environment are immutable, and they have inherent security of blockchain. The latency for new transactions and data retrieval is observed. In the future, we intend to provide more implementation details and experimental results with improvements in the scope of this work.

References

[1] Dr. M. L. Ravi Chandra, B. Varun Kumar and B. Sureshbabu. (2017). Smart Home Automation Using Virtue of IoT. IEEE International Conference on Energy, Communication, Data Analytics and Soft Computing, P1–P5.

[2] Ioan Szilagyi and Patrice Wira. (2018). An Intelligent System for Smart Buildings Using Machine Learning and Semantic Technologies: A Hybrid Data-Knowledge Approach. *2018 IEEE Industrial Cyber-Physical Systems (ICPS)*, St. Petersburg, Russia, P20–P25.

[3] Yuanyu Zhang, Shoji Kasahara, Yulong Shen, Xiaohong Jiang and Jianxiong Wan. (2018). Smart Contract-Based Access Control for the Internet of Things. *IEEE*, 6(2), P1–P11.

[4] Konstantinos Christidis and Michael Devetsikiotis. (2016). Blockchains and Smart Contracts for the Internet of Things. *IEEE Access*, 4, P1–P12.

[5] Kamanashis Biswas and Vallipuram Muthukkumarasamy. (2016). Securing Smart Cities Using Blockchain Technology. IEEE International Conference on High Performance Computing and Communications, P1–P2.

[6] Waqar Ali, Ghulam Dustgeer, Muhammad Awais, and Munam Ali Shah. (2017). IoT Based Smart Home: Security Challenges, Security Requirements and Solutions. International Conference on Automation & Computing, University of Huddersfield, P1–P6.

[7] Vaibhavkumar Yadav, Shubham Borate, Soniya Devar, Rohit Gaikwad and A. B. Gavali. (2017). Smart Home Automation Using Virtue of IoT. IEEE International Conference for Convergence in Technology, P1–P5.

[8] Ms. Priti Vasant Kale, Dr. Samidha Dwivedi Sharma. (2014). Intelligent Home Security System Using Illumination Sensitive Background Model. *International Journal of Advance Engineering and Research Development*, 1(5), P1–P11.

[9] Arun Cyril Jose, and Reza Malekian. (2017). *Improving Smart Home Security; Integrating Logical Sensing into Smart Home*. IEEE, P1–P18.

[10] Dariusz Frejlichowski, Katarzyna Gosciewska, Paweł Forczmanski and Radosław Hofman. (2014). "Smartmonitor"—An Intelligent Security System for the Protection of Individuals and Small Properties with the Possibility of Home Automation. *Sensors*, 14(6), P1–P27.

[11] Timothy Malche and Priti Maheshwary. (2017). Internet of Things (IoT) for Building Smart Home System. IEEE International Conference on I-SMAC (IoT In Social, Mobile, Analytics and Cloud) (I-SMAC), P1–P6.

[12] S. Tanwar, P. Pately, K. Patelz, S. Tyagix, N. Kumar and M. S. Obaidat. (2017). An Advanced Internet of Thing Based Security Alert System for Smart Home. IEEE International Conference on Computer, Information and Telecommunication Systems, P1–P5.

[13] Dr. M. L. Ravi Chandra, B. Varun Kumar, B. Sureshbabu. (2017). IoT Enabled Home with Smart Security. International Conference on Energy, Communication, Data Analytics and Soft Computing, P1–P5.

[14] Joshua Streiff, Olivia Kenny, Sanchari Das, Andrew Leeth, and L. Jean Camp. (2018) Poster Abstract: Who's Watching Your Child Exploring Home Security Risks with Smart Toybears. IEEE/ACM Third International Conference on Internet-of-Things Design and Implementation, P1–P2.

[15] Su Zin Zin Win, Zaw Min Minhtun, and Hlamyo Tun. (2016). Smart Security System for Home Appliances Control Based on Internet of Things. *International Journal of Scientific & Technology Research*, 5(6), P1–P6.

[16] Ali Dorri, Salil S. Kanhere, Raja Jurdaky and Praveen Gauravaram. (2017). Blockchain for IoT Security and Privacy: The Case Study of a Smart Home. IEEE International Conference on Pervasive Computing and Communications Workshops (Percom Workshops), P1–P6.

[17] Ali Dorri, Salil S. Kanhere and Raja Jurdak. (2017). Towards an Optimized Blockchain for IoT. ACM Second International Conference on Internet-of-Things Design and Implementation, P1–P6.

[18] Ali Dorri, Salil S. Kanhere, Raja Jurdak, and Praveen Gauravaram. (2017) *A Light Weight Scalable Blockchain for IoT Security and Privacy.* IEEE, P1–P17.

[19] Raja Jurdak. (2017). *Blockchain for Internet of Things Security and Privacy.* Csiro, P1–P21.

[20] Ali Dorri, Salil S. Kanhere, Raja Jurdak, Praveen Gauravaram. (2017). *Blockchain for IoT Security and Privacy.* IEEE, P1–P7.

[21] Jianjun Sun, Jiaqi Yan and Kem Z. K. Zhang. (2016). Blockchain-Based Sharing Services: What Blockchain Technology Can Contribute to Smart Cities. *Financial Innovation*, 2, P1–P9.

[22] Shiyong Yin, Jinsong Bao, Yiming Zhang and Xiaodi Huang. (2017). M2M Security Technology of CPS Based on Blockchains. *Symmetry*, 9(9), P1–P16.

6

IoT-Based Robotic Arm

VINAY KUMAR AWAAR
AND PRAVEEN JUGGE

Gokaraju Rangaraju Institute of Engineering and Technology, Hyderabad, India

Contents

6.1 Introduction

Robots are capable of amazing feats of strength, speed, and seemingly intelligent decisions; however, this last ability is entirely dependent upon the continuing development of machine intelligence and logical routines [1]. Our future is closely bound with robotics as it is becoming an intrinsic part of the human race. The International Federation of Robotics (IFR) defines a service robot as a robot that operates semi- or fully autonomously to perform services useful to the well-being of humans and equipment, excluding manufacturing

DOI: 10.1201/9781003269991-6

operations [2]. The new advancements in the field of robotics are giving us an unprecedented definition of what robots can do.

In effect, robotic arms are emerging in automobile and manufacturing industries, automating the process and thus streamlining it. There are many companies offering robotic arm models, but teaching how to operate and change the functionality or position of the robotic arm to a factory worker who is a layman can become a painstakingly arduous task. Thus, in this paper we propose an Android app enabled IoT-based robotic arm with weight-based object segregation that even a layman can control with their fingertips. Though there are papers on robotic arms and Bluetooth-enabled robotic arms, our attempt in this paper is to develop an Android app enabled Bluetooth-controlled robotic arm with a weight-based object segregation.

The weight-detecting sensor decoupled with the robotic arm will detect and display the weight of the object picked up by the robotic arm. This feature can help the manufacturing and packing industries where sometimes the robotic arm can pick up weight exceeding its limit and can lead to malfunction or breaking of the robotic arm. This can also help in packaging in the manufacturing industry, where the weight of the product needs to be measured and segregated before packing it.

The robotic arm system is designed by using components and hardware with embedded software to provide an Android app–controlled robotic arm, based on Bluetooth technology, to automate the manufacturing industry. The mass of the object is calculated using a custom-built 50 kg load sensor module and is then exhibited on a serial monitor in real time along with the tag number, so that it can be logged and tracked accordingly; the objects picked up by the robotic arm can then be segregated based on their weight.

6.2 Literature Review

In the past, some researchers proposed various means of designing, building, and controlling a robot arm, for instance, a previous project on autonomous robot navigation using radio frequency (RF) [3]. The robot was prepared mechanically to be suitable for this RF to work. In a later published work, researchers Kurt E. Clothier and Ying Shang proposed "A Geometric Approach for Robotic Arm Kinematics with Hardware Design, Electrical Design, and

Implementation" [4], where they initiated a dimensional perspective to the robotic arm.

With time, there have been multiple papers published on the robotic arm that also include "Robotic Arm Control using Bluetooth Device" with an Android application [5], in which they put forward a mechanism to control the robotic arm using a Bluetooth module, which is close to our project in terms of implementation. What we are proposing in this paper is a more efficient and robust variant of the robotic arm that can revolutionize the manufacturing industry.

6.3 Methodology

6.3.1 3D Modeling Analysis of the Robotic Arm

The robotic arm was designed using Solid Works software; the arm includes five axes for the movement of different joints. The first three axes, which are the elbow, shoulder, and the waist, are equipped with MG996R servo. The other two axes are joints, which are the wrist pitch, and the wrist roll is equipped with SG 90 servos (Figure 6.1).

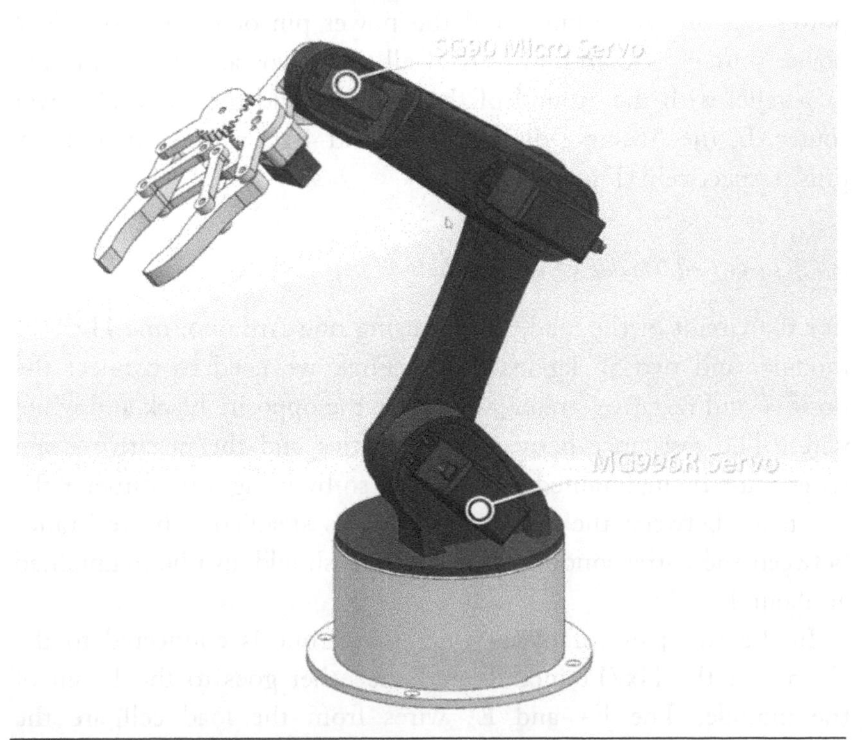

Figure 6.1 3D model of a robotic arm.

The gripper, which is used for the pick and place function, is also equipped with a SG90 servo. The parts of the arm were 3D printed and assembled using 2 mm and 4 mm screws.

6.4 Electrical Circuit Analysis

6.4.1 Robotic Arm Circuit Analysis

For the circuit of the robotic arm, we are using the Arduino uno and the HC-05 module for the communication. We are using the servos SG90 and MG996R as actuators. The servo motor is one of the DC-type motors with feedback that is used in many applications that require controlling the system in an up-down direction. Servos are extremely useful in robotics [6]. As we can see, the six digital pins of the Arduino are attached to six servo motors. Also, we need an ambient power source of voltage, 5 V, and current, 2 amps, to power the servos as the power from the Arduino isn't sufficient to move the servos.

The power pins of all the servos are connected in parallel with the power pin of the Arduino and the power pin of the external 5 V power source. The ground pins of all the servos are also connected in parallel with the ground of the Arduino and the external power source. In the Arduino, digital pins 3 and 4 act as the TX and RX pins, respectively (Figure 6.2).

6.4.2 Load Cell Module Circuit Analysis

For the circuit of the load, we are using one Arduino, one HX 711 module, and two 50 kg load cells. First, we need to connect the positive and negative strains, which are the opposite black and white wires. The resistance between the positive and the negative strain wires must be maintained at 1 k ohm, so by using a multimeter the resistance between the set of two wires is steadied. The resistance between the corresponding two red wires should also be maintained at about 1 k ohm.

In the two pairs of black-white wires, one is connected to the E+ pin of the Hx711 module and the other goes to the E-pin of the module. The E+ and E- wires from the load cell are the power wires. The red wires are connected to the A+ and A- pins of

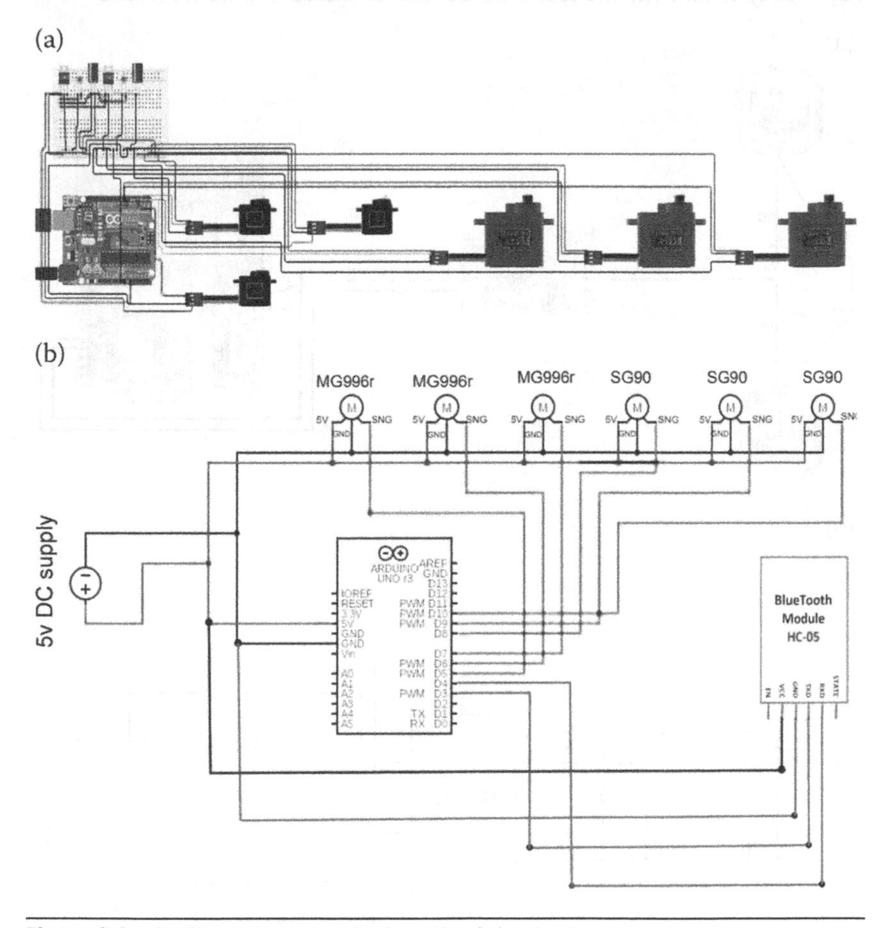

Figure 6.2 (a) Circuit diagram; (b) schematic of the circuit.

the Hx711 module, which act as the measurement input pins. By connecting the DT and SCK pins to digital pins 4 and 5, VCC to 5 V, and GND to the ground of Arduino, the circuit of the weight sensor is complete (Figure 6.3).

The two-load cell weight sensor circuit is coupled to the bottom of the robotic arm so that it computes the weight of the objects that are picked up by the arm. As the load cell will give the output in microvolts and Arduino is not capable of reading these values, amplification is required. The best solution is to use a HX711 amplifier, which is a 24-bit analog-to-digital amplifier [7]. While calculating the weight of the object, we have calibrated the sensor in such a way that it considers the weight of the robotic arm as tare and displays the value.

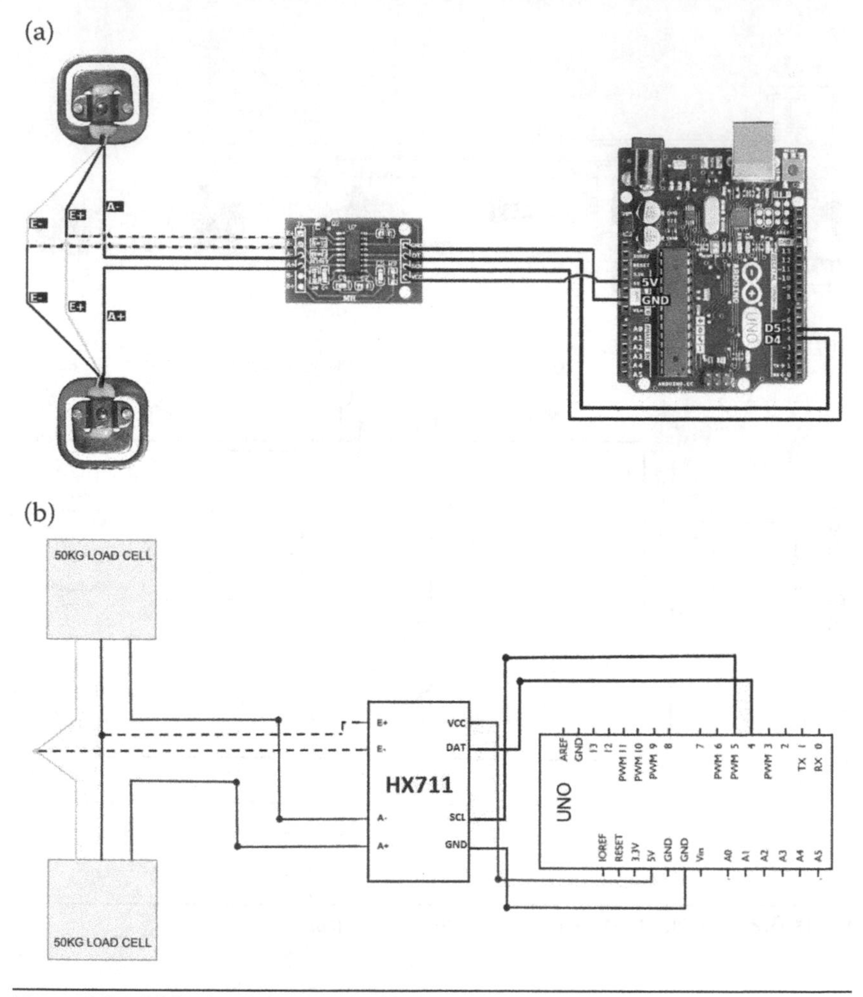

Figure 6.3 (a) Load cell circuit model; (b) load cell module circuit schematic.

6.5 Graphical User Interface Analysis

The MIT App Inventor is a drag-and-drop interface visual programming tool that allows everyone to design and build fully functional mobile apps for Android. App Inventor advocates a new era of personal mobile computing in which people are able to design, create, and employ personally relevant and meaningful mobile technology solutions for a variety of contexts in their daily lives, in endlessly unique situations [8].

The input to the servos is given through the app via the Bluetooth module, and the app was built using the MIT app inventor and interface, which can be seen in the image. Each axis servo has a separate control to it on the app's display along with the arm speed that can also be controlled through the app.

We have used an Android application that was designed using the MIT Application Inventor to control and operate the robotic arm in various modes of speeds and angles. This robotic arm is mainly controlled using the Android device with the help of the Bluetooth module that is present in the internal circuit of the robotic arm. By viewing at the site of the Android application and the Bluetooth module, it can be known which data are actually fleeting to the Arduino (Figure 6.4).

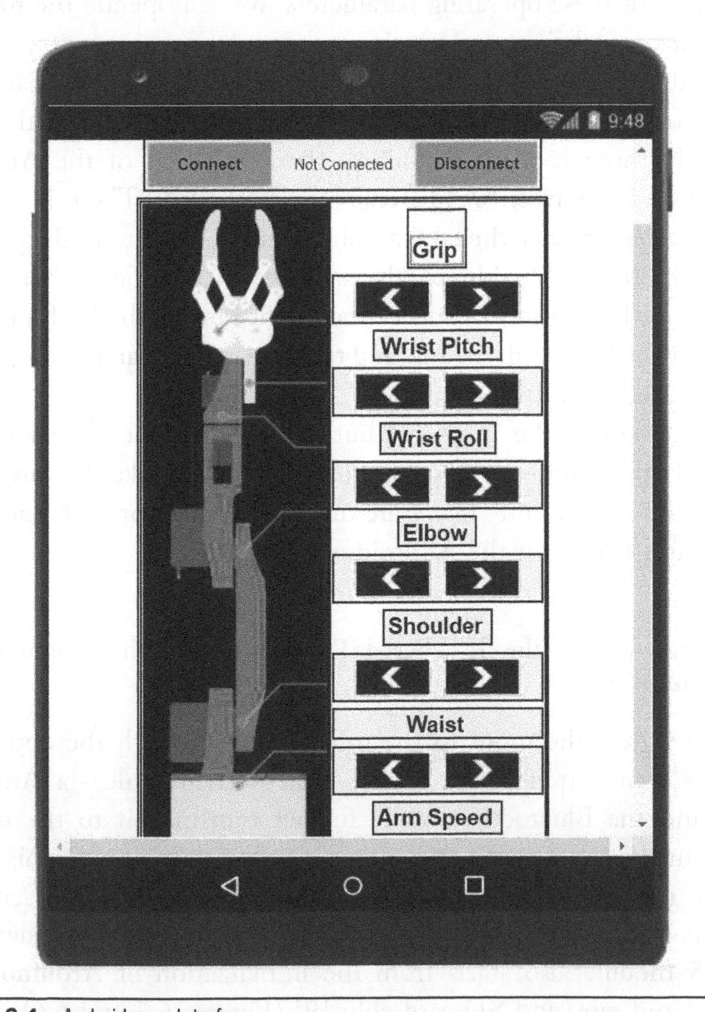

Figure 6.4 Android app Interface.

When seen at the Android application interface at the beginning, there are two buttons for connecting the smartphone to the HC-05 Bluetooth Module via the created Android application. Next, on the left side there is an image preview of the robotic arm that shows all the different joints in the robotic arm. Then, on the opposite to this robotic arm preview, there is a slider tab for easy control and movement of the servos and one more slider for various speed control. When it comes to control and operation of the robotic arm for moving the joints, there are different options, and they are grip control, wrist pitch control, wrist roll control, elbow control, shoulder control, and waist control and finally arm speed control. By changing the values of these operating parameters, we can operate the robotic arm in a very efficient and precise manner.

The slider tabs have a variant starting, low and high value, that takes care of the robotic arm joints. The robotic arm can be controlled to run inevitably through three buttons at the conclusion of the Android application's interface: SAVE, RUN, and RESET. There is also an option to demonstrate the total number of steps taken until the present stage. At the right side of the application interface, there is a Disconnect button to disconnect the connection with the Bluetooth module. The Bluetooth module and the smartphone can be completely disconnected with this button.

The design of the interface, buttons, and all the sliders in the Android application are possible only with the blocks that are built behind the application. There are different blocks for each specified option and feature in the Android application.

6.6 Mechanism of the IoT-Based Robotic Arm with Weight-Based Segregation

The user gives the input to the robotic arm through the app. The input is transmitted to the Hc-05 Bluetooth module via Arduino Uno, and the Bluetooth module further transmits it to the servos placed in the robotic arm to perform a certain function, in this case, to pick up or place an object of interest. The process of sending requests and receiving a response of sensed data using the Bluetooth HC-05 module also starts from the initialization of Arduino pin, status, baud rate, and SD card chip [9] (Figures 6.5 and 6.6).

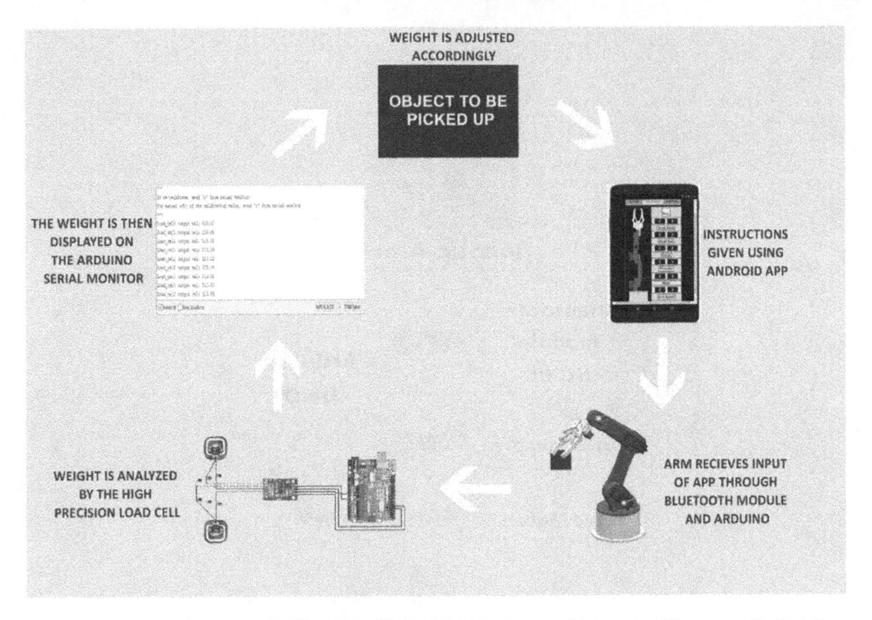

Figure 6.5 Process chart of the weight-based segregation using 5 DOF robotic arm.

Once the object is picked up, the weight of the object is calibrated by the load cells. Since they are force transducers, they convert the weight or pressure or force into electrical current. This conversion and transmission to the serial monitor is done by a HX711 module and Arduino. On the serial monitor screen, the values are displayed along with the time stamp.

6.7 Simulation Results

As we can infer from the image, the weight of the object can be seen displayed on the screen at different time stamps. From these time stamps, we can also calculate the weight of the object as the arm is moving while different objects are being lifted by the arm.

The weight of the robotic arm is considered as tare in order to avoid the arm's weight adding to the picked-up object's weight. Tare basically means that the sensor is calibrated in such a way that it doesn't consider the weight of the robotic arm in the process of lifting the object; it is irrelevant to the angle or position in which the object is being lifted.

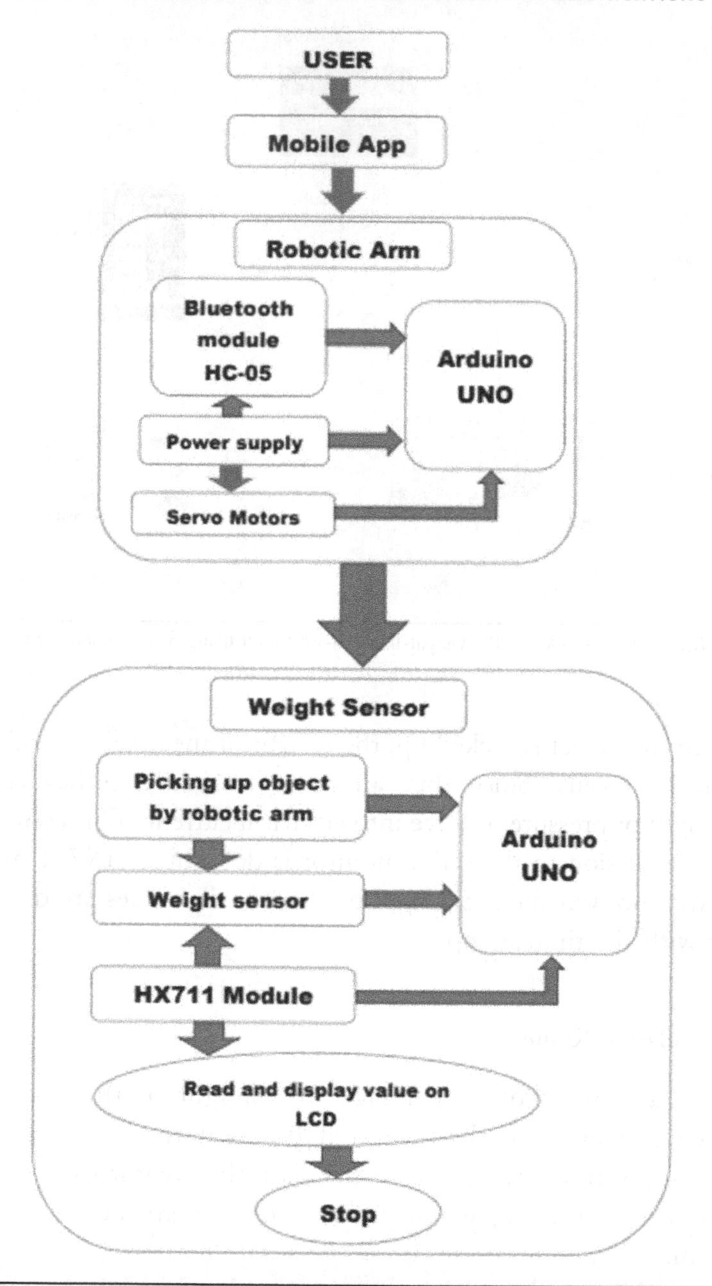

Figure 6.6 Flow chart of the workings of the robotic arm with a load cell module.

As we can see from Figure 6.7, we can infer that no object is being picked up by the arm; hence, the value being displayed on the serial monitor is close to zero. From this we can also conclude that the weight of the arm is being considered as tare, whereas in Figure 6.8

```
16:29:07.696 -> Load_cell output val: 0.65
16:29:07.790 -> Load_cell output val: 0.64
16:29:07.883 -> Load_cell output val: 0.64
16:29:07.976 -> Load_cell output val: 0.64
16:29:08.068 -> Load_cell output val: 0.65
16:29:08.160 -> Load_cell output val: 0.67
16:29:08.252 -> Load_cell output val: 0.69
16:29:08.345 -> Load_cell output val: 0.70
16:29:08.439 -> Load_cell output val: 0.69
16:29:08.532 -> Load_cell output val: 0.69
16:29:08.625 -> Load_cell output val: 0.66
16:29:08.717 -> Load_cell output val: 0.63
16:29:08.809 -> Load_cell output val: 0.64
```

Figure 6.7 Weight displayed on the serial monitor when there is no load.

```
***
To re-calibrate, send 'r' from serial monitor.
For manual edit of the calibration value, send 'c' from serial monitor.
***
Load_cell output val: 113.03
Load_cell output val: 113.06
Load_cell output val: 113.08
Load_cell output val: 113.09
Load_cell output val: 113.12
Load_cell output val: 113.14
Load_cell output val: 113.12
Load_cell output val: 113.08
Load_cell output val: 113.09
Autoscroll  Show timestamp                Both NL & CR  57600 baud
```

Figure 6.8 Weight displayed on the serial monitor when there is a load.

we can see that there is a difference in the value being displayed on the monitor, which means that the arm is holding an object.

This result can be useful in segregating the objects based on their weight and monitoring the load the arm is being administered so that if we place a larger load, which can result in damage to the servos. We can intimate this method is also being researched for use in injection mold–based product manufacturing, where the weight of the object to be measured is very important to maintain the precision in molding and weight of the whole product.

6.8 Result

In this way, we have implemented an IoT-based robotic arm with weight-based segregation controlled wirelessly by a graphical user interface. Therefore, the robotic arm is successfully connected with the smart device and with the Bluetooth module, and this total setup was further ordered to perform the segregation based on the parameter "WEIGHT". We can see that the load cell is working as per the given instructions. It is given the task of displaying the weight quantity of the load the robotic arm picks. And then we can see that the robotic arm places the object at the specified place without any deviation.

We can see that the robotic arm accepts all the commands that the user passes through the Android application. The load cell performs and fulfills all the tasks entrusted to it without hesitation. At the next step, the load cell's readings are mirrored on the serial monitor. By all these tasks and after the appropriate testing of the load cell and weight sensor, it can be stated that the weight-based segregation using the IoT-based robotic arm is carried out precisely. As a sign of this result, we have administered various commands to the arm by picking and placing different objects and monitoring and segregating the load being apportioned to the arm.

This method helps immensely in inline weight measurement without the need of picking up, weighing, and placing the object on the conveyor line every time. It has eliminated the need for weighing each object separately after production. Inline weighing can act as a 100% parts check because on the basis of weight, plastic processors are able to ascertain if the sprues (a channel through which metal or plastic is poured into a mold) are cut clean or whether excessive or insufficient injection has been done.

And after we have performed the weight segregation with the help of the robotic arm and the load cell sensor by taking numerous loads with variant weights, we can determine that the robotic arm is working efficiently by weighing all the different loads precisely and by also determining missing parts in two identical objects with the help of weight. The results can be illustrated in Figure 6.9.

6.9 Conclusion

In the present scenario, robotic arms are taking over the workload of humans by simplifying and automating various processes. Although significant development in this field is necessary, from

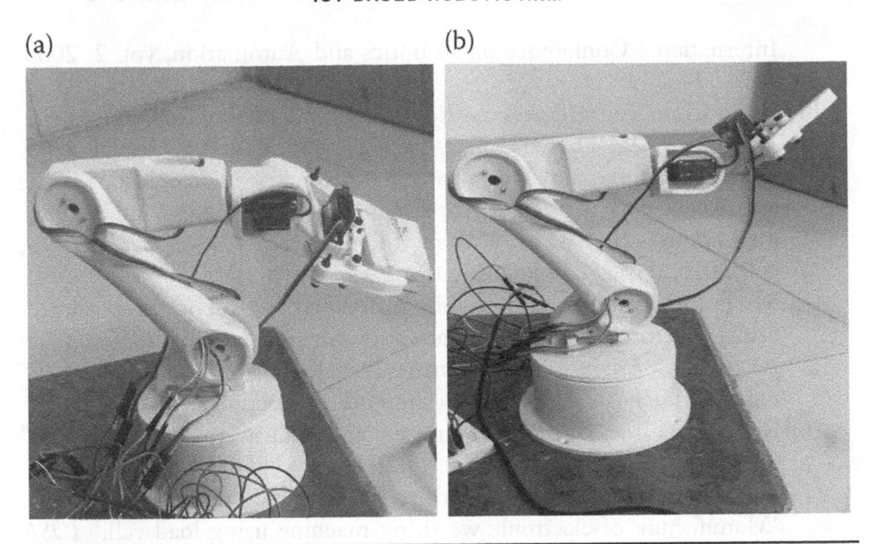

Figure 6.9 (a) Final resulting product of robotic arm with load cell module; (b) final resulting product of robotic arm with load cell module lifting a load.

its inception, the amount of human error and involvement has been significantly reduced.

The purpose of our project is to provide control of a five-axis moving robot arm design. This robot arm, with a suitable microcontroller and Bluetooth module with an Android application, can also monitor the load and segregate objects based on their weight. And this cutting-edge technology is really helpful in industries with injection molding where the component weight demonstrates the quality of the component and also the process consistency. With the aid of weight-based segregation, the NOK, or not okay, parts can be easily separated from the good ones. The necessary theoretical and practical information for this purpose has been obtained, and the necessary infrastructure has been established for the project.

References

[1] C. C. Kemp, A. Edsinger, and E. Torres-Jara, "Challenges for robot manipulation in human environments [Grand challenges of robotics]," *IEEE Robotics and Automation Magazine*, vol. 14, no. 1, pp. 20–29, 2007.

[2] R. C. Luo, and K. L. Su, "A multi agent multi sensor based real-time sensory control system for intelligent security robot," IEEE

International Conference on Robotics and Automation, vol. 2, 2003, pp. 2394–2399.

[3] Ming Chun Tan, *Autonomous Robot Navigation Using Radio Frequency. Bachelor Project*. Thesis. University Teknologi Malaysia, Skudai, 2005.

[4] Kurt E. Clothier, and Ying Shang, "A geometric approach for robotic arm kinematics with hardware design, electrical design, and implementation," *Hindawi Publishing Corporation Journal of Robotics*, vol. 2010, Article ID 984823, 10 pages.

[5] M. Pon Alagappan, and N. Shivaani Varsha, "Robotic arm control using bluetooth device with an android application," *International Journal of Engineering Research & Technology (IJERT) IJERT* www.ijert.org NCACCT'14 Conference Proceedings.

[6] Riazollah Firoozian, "Servo motors and industrial control theory," Springer, 1st edition (December 8, 2008).

[7] Snehashis Das, Avijit Karmakar, Pikan Das, and Biman Koley, "Manufacture of electronic weighing machine using load cell," *IOSR Journal of Electrical and Electronics Engineering (IOSR-JEEE) e-ISSN: 2278-1676, p-ISSN: 2320-3331*, vol. 14, Issue 4 Ser. I, pp. 32–37, July– August 2019.

[8] Shaileen Crawford Pokress, and Jose Juan Dominguez Veiga, "MIT app inventor enabling personal mobile computing." *arXiv preprint arXiv:1310.2830.*

[9] Mahar Faiqurahman, Diyan Anggraini Novitasari, Zamah Sari, and Universitas Muhammadiyah Malang, "QoS analysis of kinematic effects for bluetooth HC-05 and NRF24L01 communication modules on WBAN system," *KINETIK*, vol. 4, No. 2, pp. 187–196, May 2019.

7

Assimilation of Blockchain for Augmenting the Security and Coziness of an IoT-Based Smart Home

C.H. NAGARAJU[1], BANDI DOSS[2], P. BALAMURALIKRISHNA[3], D. LAKSHMAIAH[4], AND I. VENU[5]

[1]*Professor of ECE Department, AITS, Rajampet, India*
[2]*Professor of ECE Department, CMR Technical Campus, Hyderabad, India*
[3]*Dean R&D and Professor of EEE Department, Chaplapthi Institute of Technology, Guntur, India*
[4]*Professor of ECE Department, SriIndugroups, Hyderabad, India*
[5]*Assistant Professor of ECE Department, Sri Indu Institute of Engineering and Technology, Hyderabad, India*

Contents

7.1 Introduction

In all fields, including in network fields, technology has improved. User technology has changed to the P2P network. In process

DOI: 10.1201/9781003269991-7

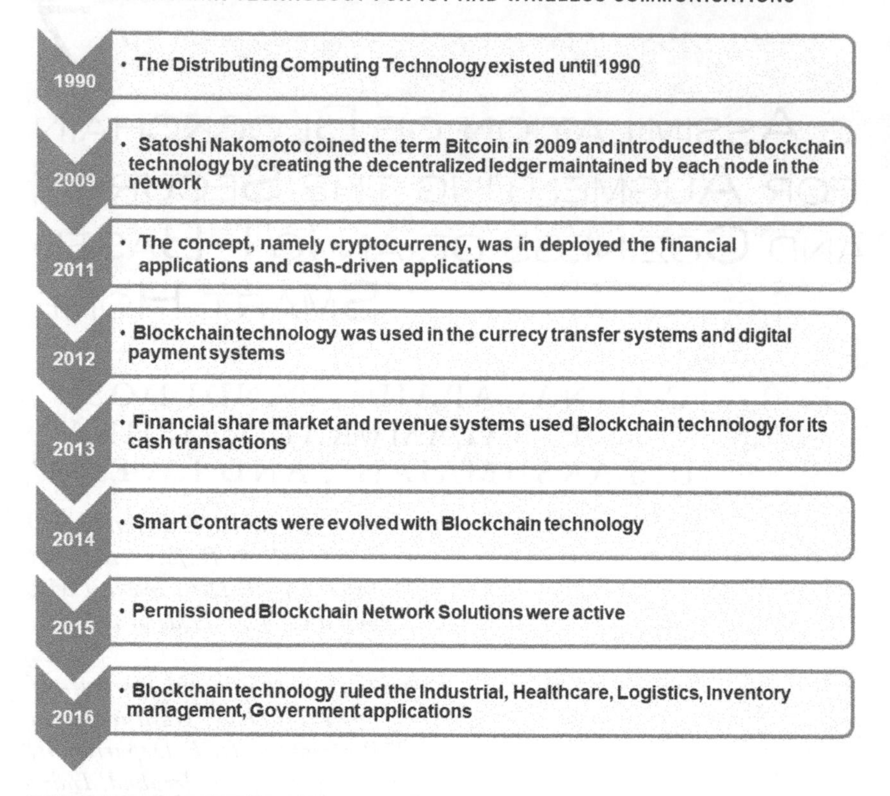

1990
• The Distributing Computing Technology existed until 1990
2009
• Satoshi Nakomoto coined the term Bitcoin in 2009 and introduced the blockchain technology by creating the decentralized ledger maintained by each node in the network
2011
• The concept, namely cryptocurrency, was in deployed the financial applications and cash-driven applications
2012
• Blockchain technology was used in the currecy transfer systems and digital payment systems
2013
• Financial share market and revenue systems used Blockchain technology for its cash transactions
2014
• Smart Contracts were evolved with Blockchain technology
2015
• Permissioned Blockchain Network Solutions were active
2016
• Blockchain technology ruled the Industrial, Healthcare, Logistics, Inventory management, Government applications

Figure 7.1 Introduction of blockchain technology.

advancement, it has evolved in WAN and LAN with network portability. So it emerged as World Wide Web network technology. In parallel processing, again the network divided into central, decentralized, and distributed area networks. Any iterations can enter into all open forums and update their transactions in all blockchain systems [1–5] (Figure 7.1).

Online applications such as e-trade, online shopping, etc. can have the advantage of blockchain technology for perfecting their functional effectiveness. The blockchain can be either public or private. Any stranger or unknown user can join and sign up his/herself inside the blockchain and process the online transaction. It is more transparent in execution to all their users because of nodes; guests are also registered. Certified iterations are securely logged in to the network. Authorized transactions are accepted authenticated, and stored in the database as a distribution list. For a specified time period, all the users' distribution lists can be traced and executed in the process. Continuous updates happen on a regular interval, and nodes are published in a distributed ledger [6–10].

7.2 Preliminaries

It is hard for hackers to overwhelm the exchange rules to increase the transactions in the network. The agreement algorithm is used to predict the entry into the network.

7.3 Proposed Work

Sensors collect the information, and users get the notification. In smart home care, consider the feature and its nodes. The home has various nodes (Figures 7.2 and 7.3).

7.3.1 Blockchain Process

The nodes ground and receive the assignment and information from the respective nodes in the sensors (Figures 7.4 and 7.5).

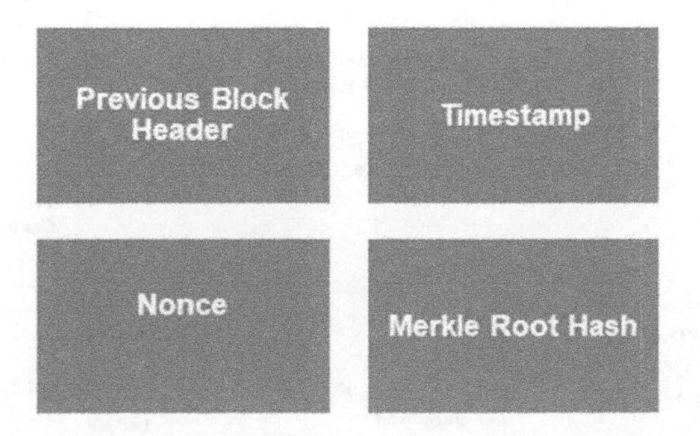

Figure 7.2 Blockchain technology process 1.

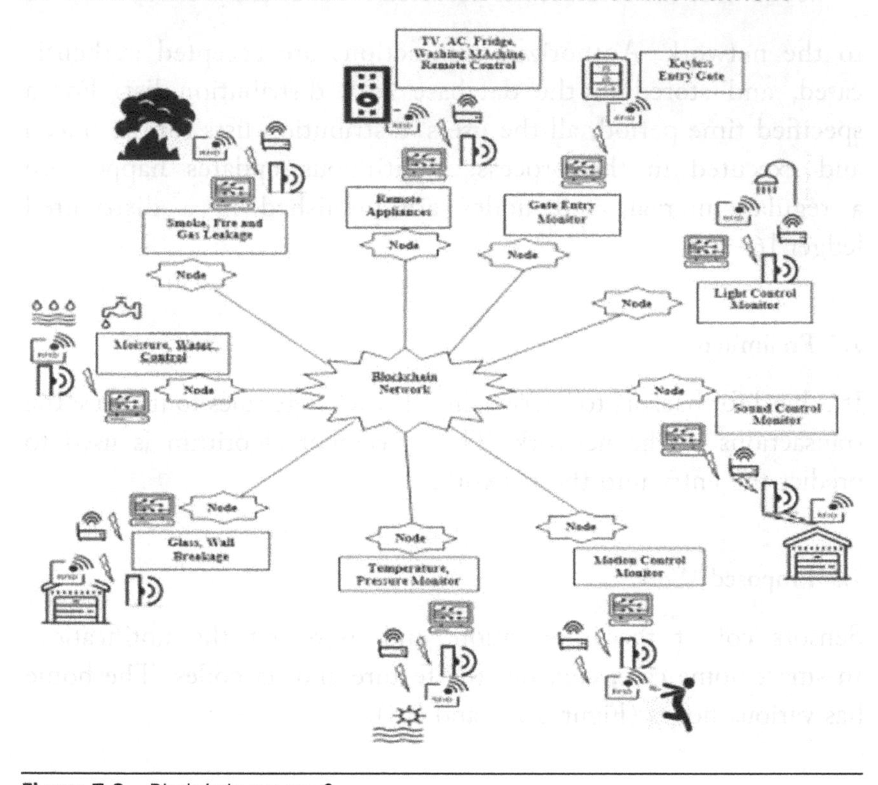

Figure 7.3 Blockchain process 2.

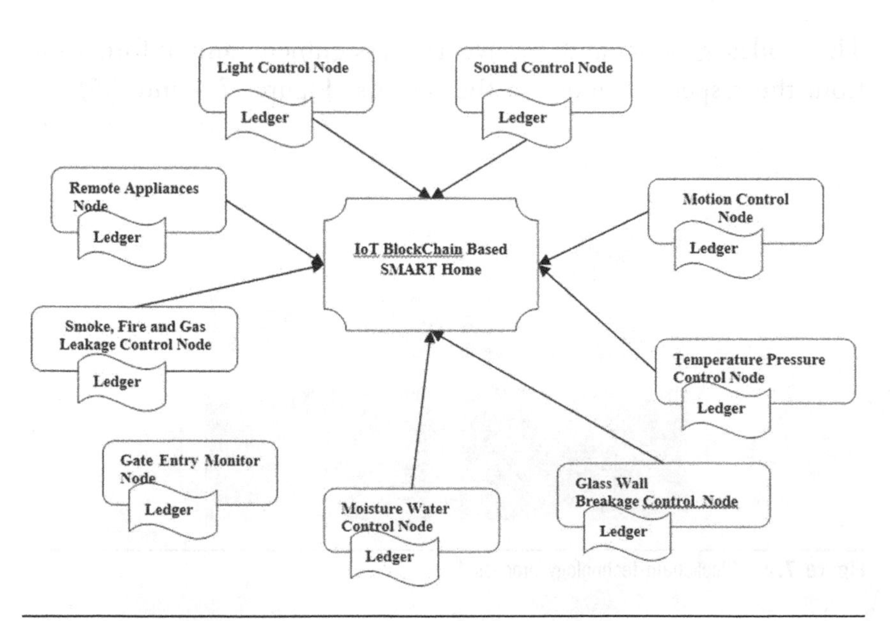

Figure 7.4 Blockchain nodes.

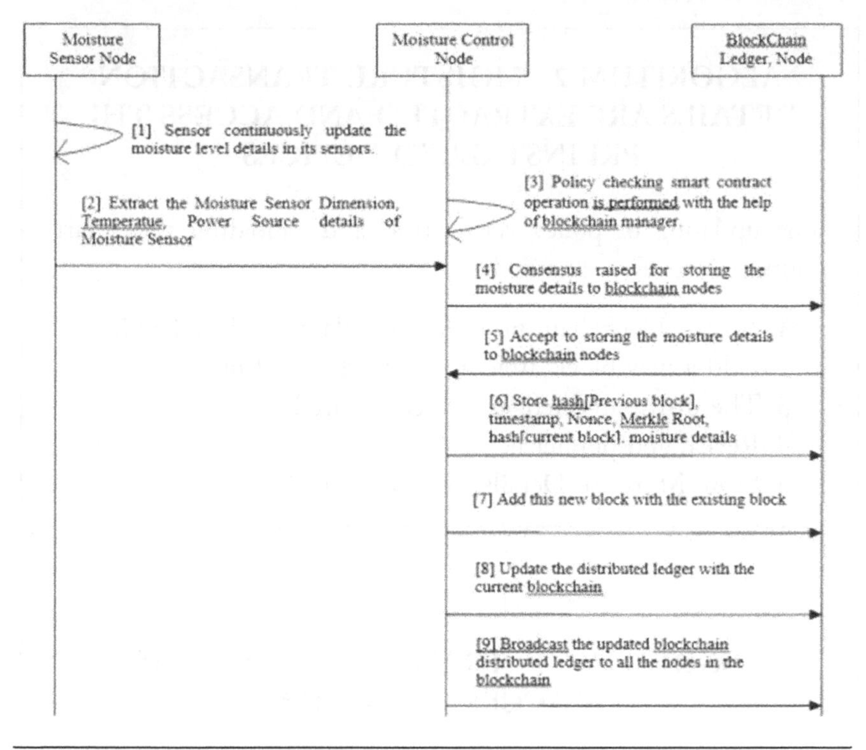

Figure 7.5 Access moisture water details use case diagram.

The algorithms are described as follows:

ALGORITHM 1

1. At the structure level, the sensors continuously update.
2. The moisture sensor is used to extract the dimension, temperature, and power source details and receive data from the internet gate ways.
3. The information is retrieved and stored.

The Moisture Water Control Node stores the moisture transaction details, monitoring and accessing them.

ALGORITHM 2 MOISTURE TRANSACTION DETAILS ARE EXTRACTED AND ACCESS THE PREINSTALLED DEVICES

The updating of policy verification and validating the smart contracts are often executed.

1. All the blockchain nodes are regulated and updated.
2. Add a new block attachment to the existing.
3. The distributed ledger is reconfigured.
4. Reveal and publicize.
5. Store Moisture Details Transaction.

ALGORITHM 3 TRANSACTION DETAILS OF ACCESS MOISTURE

1. Acquire the moisture details such as sensor dimensions, temperature, and power source.
2. Policy verification and validation are helpful in updating the smart contracts.
3. Moisture details are updated and validated.
4. Attach the new block.
5. Distributed ledger is reconfigured.
6. Reveal and publicize for each and every node.
7. Finally, store it.

ALGORITHM 4

1. Repeat and continuous observation.
2. In the predefined frequent intervals, moisture details are updated.

3. For the purpose of moisture details updates and policy verification and validation, the smart contract software is executed.
4. The new moisture control node details are retrieved after terminating the information from the required time interval.
5. All the nodes regulate and update.
6. Validate and update.
7. Attachment of the new blockchain in the existing network.
8. Reconfigure.
9. Revealing and publicizing.
10. End.

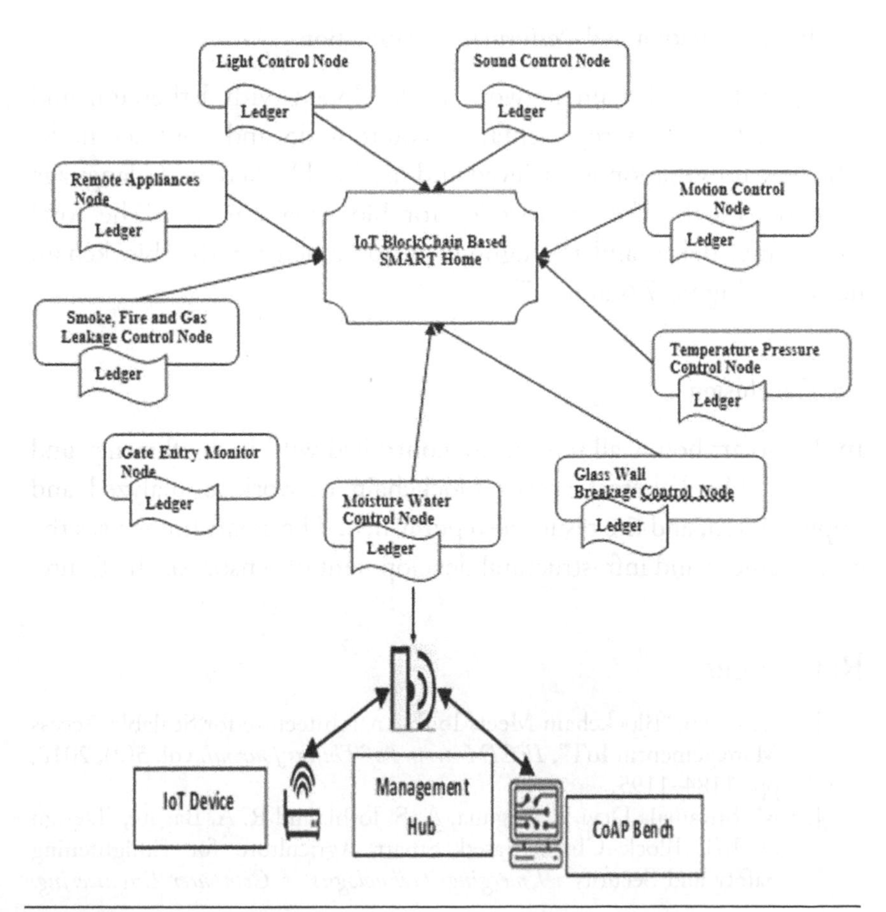

Figure 7.6 Blockchain network.

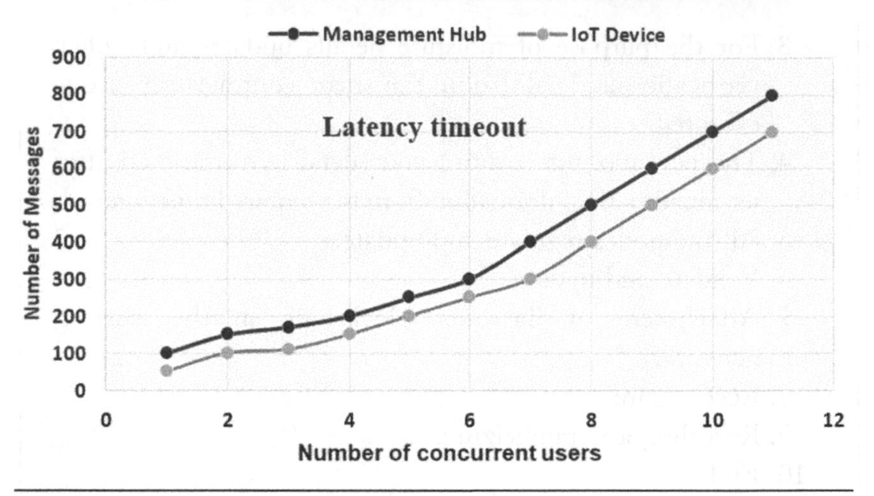

Figure 7.7 Timeout delay of moisture control node.

7.4 Implementation and Performance Evaluation

The private blockchain network is developed with Ethereum and in association with representative count node and contract node. The new information is retrieved and used in blockchain. C language is used to write the macrocodes for blockchain nodes. The total work time, delay, and throughput are obtained for this blockchain network (Figure 7.6 and 7.7).

7.5 Conclusion

In this smart home, all sensors are controlled with high efficiency and security. The Ethereal private blockchain network is analyzed and implemented, and the results also published. The smart home uses the enhancement and infrastructural development of sensors of the future.

References

[1] O. Novo, "Blockchain Meets IoT: Anarchitecture for Scalable Access Managementin IoT", *IEEE Internet of Things Journal*, vol. 5(2), 2018, pp. 1184–1195.

[2] M. Shyamala Devi, R. Suguna, A. S. Joshi, and R. A. Bagate, "Design of IoT Block Chain Based Smart Agriculture for Enlightening Safety and Security", *Emerging Technologies in Computer Engineering:*

Microservices in Big Data Analytics. ICETCE 2019, Communications in Computer and Information Science, Springer, vol. 985, 2019, pp. 7–19.

[3] R. Suguna, M. S. Devi, and R. M. Mathew, "Exploration of Block Chain for Edifying Safety and Security in IoT Based Diamond International Trade", *International Journal of Innovative Technology and Exploring Engineering*, vol. 8, no. 8, June 2019, pp. 2324–2328.

[4] G. W. Peters, E. Panayi, and A. Chapelle, *SSRN Electronic Journal*, arXiv preprint arXiv:1508.04364, 2015.

[5] L. Lamport, R. Shostak, and M. Pease, The Byzantine generals problem. In *Concurrency: the works of leslie lamport* (pp. 203–226), 2019.

[6] G.-T. Nguyen, and K. Kim, "A Survey about Consensus Algorithms Used in Blockchain", *Journal of Information Processing Systems*, vol. 14(1), 2018, pp. 101–128.

[7] A. Kosba, A. Miller, E. Shi, Z. Wen, and C. Papamanthou, "Hawk: The Block Chain Model of Cryptography and Privacy-Preserving Smart Contracts", In: Proceedings of IEEE Symposium on Security and Privacy(SP), 2016, pp. 839–858.

[8] Z. Zheng, S. Xie, H. Dai, X. Chen, and H. Wang, "Blockchain Challenges and Opportunities: A Survey", *International Journal of web and Grid Services*, vol. 14(4), 2018.

[9] K. Habib, A. Torjusen, and W. Leister, "Security Analysis of a Patient Monitoring System for the Internet of Things in Health", In: The Seventh International Conference one Health, Tele Medicine, and Social Medicine, 2015, pp. 73–78.

[10] Z. Zheng, S. Xie, H. Dai, X. Chen, and H. Wang, "An Overview of Block Chain Technology: Architecture, Consensus, and Future Trends", IEEE International Congress on Big Data (Big Data Congress), 2017, pp. 557–564. IEEE.

8

Anti-Theft Fingerprint Security System for Motor Vehicles

K. SHASHIDHAR[1], AJAY KUMAR DHARMIREDDY[2], AND CH. MADHAVA RAO[3]

[1]*Department of ECE, Professor, Lords Institute of Engineering and Technology, Hydrabad, India*
[2]*Department of ECE, Asst. Professor, Sir C.Reddy College of Engineering, Eluru, India*
[3]*Department of ECE, Asst. Professor, Sir C.Reddy College of Engineering, Eluru, India*

Contents

DOI: 10.1201/9781003269991-8

8.1 Introduction

These days, even affordable mobile phones will have high-security features like fingerprint sensors. Until now, most motor vehicles did not have any security systems [1]. This might be one of the reasons for the tremendous theft of vehicles in our daily lives. In India, statistics show that motor vehicle theft has become one of the most unsolved crimes in every city. In 2020, over 493,000 theft cases were reported across India [2]. This was a significant drop in the number of cases compared to the previous fiscal years. A small innovation in the automobile sector sparks the importance of security in motor vehicles. According to our study, creative design [3] has been implemented to provide safety for automobiles by including hardware components like fingerprint sensors and GPS (Global Positioning System)/GSM (Global System Mobile Communication) modems [4]. This innovation would be the most effective extension of the security system in motor vehicles. In this project, a fingerprint sensor is outlined inexpensively for a vehicle. A GPS/GSM modem can be installed to track the coordinates of a bike's location when theft has occurred.

The GPS/GSM modem is one of the modern technologies that helps us track vehicles using a GPS module with the help of a GSM module. In this prototype, we can track the coordinates of the vehicle with smart phones through the use of Google maps, and the live location of the vehicle is updated every 15 minutes [5]. The Arduino programmes a GPS/GSM modem, which is powered by a microcontroller. The primary goal of this project is to improve the in-dash protection for motor vehicles by adding a fingerprint sensor. The user will be alerted by the GSM module if any unauthorised user attempts to place his/her finger. In case of any theft, we can use the GPS modem to track the vehicle's location. The ignition of the vehicle will be turned off by sending commands from the GSM modem.

8.2 Literature Survey

Emerging location-aware mobile technologies are successfully utilised in cultural contexts in this project. Many technologies such as RFID and Wi-Fi are used to contact mobile devices [6].

This project is extremely advantageous. When necessary, it can provide location-aware information. Users can choose the information they would like to see manually. The author proposes that if the attempt to dispose of the property is made promptly, it can reduce high-value property loss due to theft [7]. This project is about the anti-theft system that uses RFID technology. An accelerometer is built inside the RFID tag that connects objects. When it comes to defending the passenger safety of a hijacked vehicle, it has a lower hacking probability [8]. We describe an anti-theft control organisation for automobiles in this project, which aims to prevent vehicle theft. This technology [9] uses an integrated chip to activate a proximity sensor, which detects the key as it is inserted and sends an SMS to the owner's phone alerting the automobile access. The warning about unauthorised use is forwarded to the owner. This technical mechanism has lost its allure as it has become commonplace in cars for alarms to go off without reason.

This project focuses on biometrics-confidential settings for privacy leaks [1]. We request that it release information regarding biometric data as much as possible in addition to the secret. This project describes a motorcycle safety system that prevents motorcycle theft. Mobile contacts were employed in the system, which was designed using a microcontroller and a global layout. The system can be installed in a secret location on a motorcycle. If the machine is triggered without the key for the microcontroller, an act of pressing the paddle or a signal will be sent. The status will be detected by the microcontroller. It has a limit switch linked to it, which sends a notification to the owner via cell phone and quickly stops the engine. After the owner has disabled the system, it will resume normal operation. The technology is recommended as being simpler and more efficient than traditional security systems.

The author claims that advances in current science and technology have made it possible to apply biotechnology in everyday life. Fingerprint recognition is a relatively new high-tech application. The fingerprint image can now receive its characters. It can be used in a variety of physiology-related domains. Individuality and invariance are characteristics of the human fingerprint. The author claims that one of the most obvious truths in the world is the increase in the number of cars and other vehicles produced during the production era, as well as theft attempts. Many people and businesses work hard to develop car safety systems, but the results are less than ideal, and the number of car theft cases continues to climb. Thieves are honing their abilities and discovering the best and most powerful stealing methods, which necessitate the use of more sophisticated defence systems. This research project suggests a car monitoring and security model to address this issue. It introduces a sophisticated security model that can send an SMS and provide location information to the owner, who must respond immediately, especially if the car is nearby. The database has all of the relevant information about the vehicle and its owner, allowing police or security staff to monitor it using a GPS system that can connect to Google Earth or other mapping applications. The prototype's activation and test results demonstrate that it is capable of delivering an SMS to the owner in under 40 seconds. The tool that was created is a GSM-based embedded system. The tool is mounted to the vehicle's machine. To transmit a message to the owner's cell phone, the microcontroller is connected to an interface GSM modem. The main goal of this device is to prevent illegal entry to the car by inputting a secure password. Using the Global System for Mobile communication, the status of the same vehicle to the authorised person (owner) is intimated. This system is concerned with network security. The incorporation of mobile communications in an embedded system is a crucial element in this architecture. The entire system is designed on a single board.

8.3 Methodology

In our project, we have discussed an anti-theft security system by using fingerprint module technology until now. GPS/GSM has been installed in this project to track down the exact location of vehicles and also be able to communicate via messages from the user's to the vehicles and conversely.

A hardware part like a fingerprint module is implanted in our project to start or unlock the vehicle by the authorised person only. By using this system, the vehicle theft count will be reduced. If the vehicle is stolen, we can directly send the message to the user's phone by using the GSM module. The Arduino Uno controls the ignition of a vehicle by sending commands to stop the vehicle. Along with Arduino, GPS helps track vehicle coordinates in a real-time scenario. The GSM module provides data like text and location coordinates to the user.

8.3.1 Hardware Components

8.3.1.1 Arduino UNO Board One of the most common types of boards used with Arduino is known as the Arduino UNO. In this context, the Italian word "uno" refers to one. The first version of the Arduino software was given the designation "Uno" when it was initially distributed. Additionally, it was Arduino's first board to have a USB port when it was introduced. It is well known as a robust board that is put to use in a variety of tasks. The Arduino UNO boards were created by the company Arduino.cc. The ATmega328P is the central component of the Arduino UNO board (Figure 8.1). In comparison to certain other devices, for instance, the Mega board of Arduino, etcetera, it has a straightforward interface and is

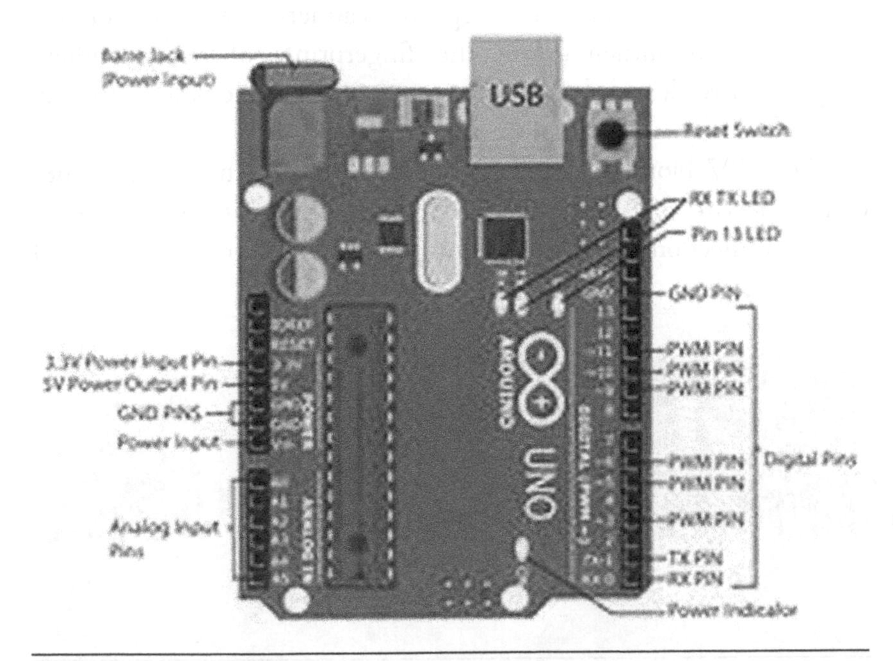

Figure 8.1 Arduino UNO board.

simple to use. Input/output pins (I/O), shields, as well as other circuitry, make up the board, which can handle both digital and analogue data. The Arduino UNO has a connection, a power jack, an ICSP (In-Circuit Serial Programming) header, and six analogue pin inputs in addition to 14 digital pins. The acronym IDE, which stands for "Integrated Development Environment," was used in the programming of this system. It is able to function properly in both online and offline environments.

8.3.1.2 RF Module Utilising an RF module (radio frequency module) is both one of the simplest and most cost-effective methods to simulate broadcast transmission (Figure 8.2). Both the transmitter and the receiver in the RF module are able to function at radio frequencies, making them an essential component of the module. In actuality, the frequency at which these modules will interact with one another is going to be either 315 MHz or 434 MHz. For this project, we can utilise a 434 MHz RF transmitter-receiver pair. The RF module will be utilised for communication for distances up to 35 metres.

8.3.1.3 Fingerprint sensor Fingerprint scanners are used for recognising and authenticating the fingerprints of an individual. Fingerprint readers and scanners are safe and reliable devices for any type of security authentication (Figure 8.3).

The R307 biometric module is a fingerprint sensor that comes equipped with a TTL UART interface. This platform enables a direct connection to a microcontroller UART or to a personal

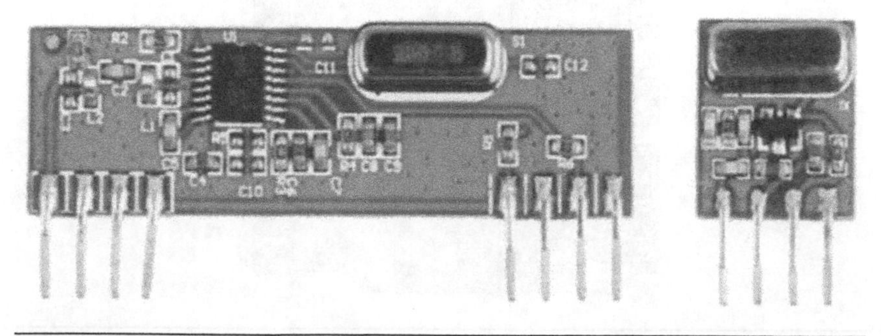

Figure 8.2 RF module.

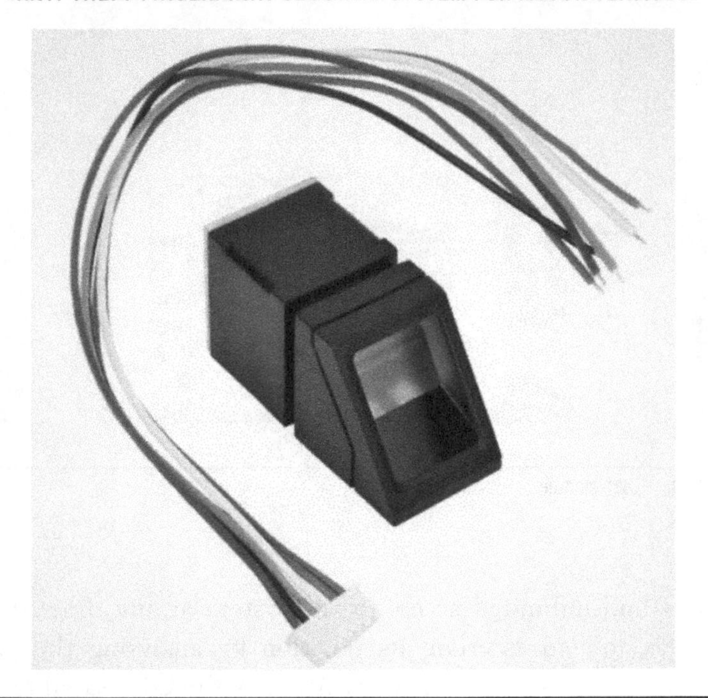

Figure 8.3 Fingerprint sensor.

computer (PC) via a MAX232/USB-Serial converter. The fingerprint data from the user can be stored in the module, and the user can choose whether to use a 1:1 or 1:N configuration for the module to identify the person.

8.3.1.4 GSM Module A GSM modem, also known as a GSM module, is a piece of hardware that utilises the technology used in GSM mobile telephones in order to establish a wireless data connection to a network (Figure 8.4). GSM modems are used in a variety of mobile devices, including mobile phones and other technology that is capable of communicating with mobile telephone networks. They identify their devices with the network via the use of SIM cards.

8.3.1.5 GPS Module The global positioning system, or more often referred to simply as GPS, is a satellite-based navigation system that provides information on both the current location and the time. It is possible for anybody who has a GPS receiver and has an unobstructed line of sight to at least four GPS satellites to

Figure 8.4 GSM module.

have free and unlimited access to the system at any time. A GPS receiver is able to ascertain its position by analysing the precise timing of the signals that are transmitted by the GPS satellites. The global positioning system (GPS) is used a lot right now and has become an important part of smartphones (Figure 8.5).

Figure 8.5 GPS module.

8.4 Proposed Design

8.4.1 Hardware Section

The circuit connections are formed by using the Arduino, fingerprint sensor, global positioning system, and accelerometer. The fingerprint sensor contains four pins, i.e., ground, supply, and one pin for transmitting the data and another one for receiving the data. The transmitter pin of the finger print sensor is connected to the transmitter pin of Arduino (pin 3), and the receiver pin is connected to the receiver pin of Arduino (pin 2). The Vcc pin is connected to a 3.3 V voltage pin, and the ground pin is grounded (Figure 8.6).

For GSM interfacing, we can make the connections the same as fingerprint sensors. To know the location of a vehicle, we can connect the Rx pin of GPS to the Arduino digital pin 5, and the Tx pin of GPS to the Arduino digital pin 4, and both GND pins are grounded.

The VCC pin of the accelerometer is connected to 5V of the Arduino Uno, the SCL (serial clock line) pin is connected to A2 of the Arduino Uno, the SDA (serial data line) pin is connected to A1 of the Arduino Uno, and both grounds are grounded. The trigger is connected to A0 of the Arduino Uno. SCL and SDA are responsible for auxiliary I2C communication. Overall connections are shown in the below figure. After completion of the

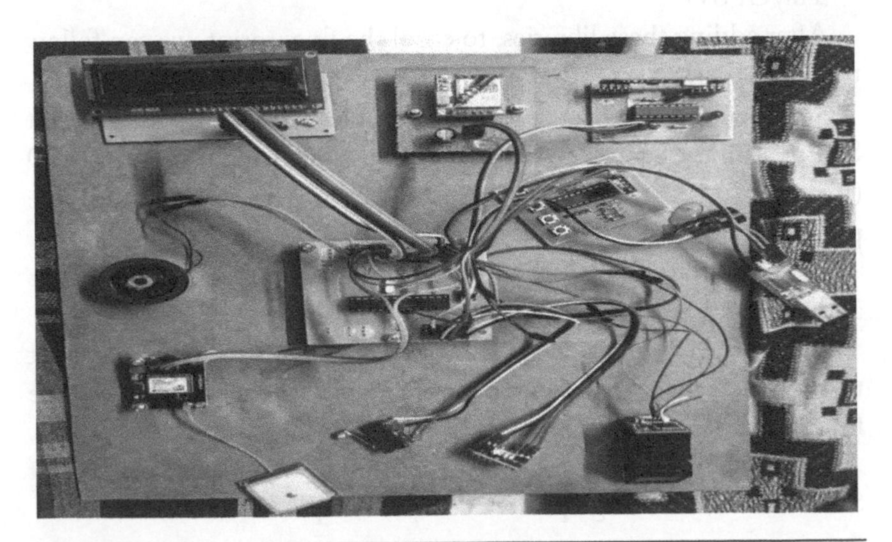

Figure 8.6 Hardware section of proposed design.

Figure 8.7 LCD display for theft detection.

hardware connections, the software section must be completed to enrol the fingerprint (Figure 8.7).

8.4.2 Software Section

To enrol and match the fingerprint, we can add some libraries as follows.

Adafruit Fingerprint Sensor Library
Unified Sensor Library by Adafruit
Library AdafruitBusIO
ADXL345 Library by Adafruit
TinyGPS
TinyGPS++

After adding these libraries, to enrol the fingerprint, we can follow the below path.

>> open the IDE -> File->Examples->Adafruit Fingerprint Sensor Library->enrol->compile->upload->open serial monitor->ID (from 1 to 127)->send->place the finger on the sensor until the process is finished

By using the above path, we can enrol the multiple fingerprints and they will be stored in the Arduino.

The following are the steps for uploading the programme to start the vehicle when an authorised person places their finger on the fingerprint sensor.

>> launch the IDE -> File->Examples->Adafruit Fingerprint Sensor Library-> Fingerprint-> compile-> upload

The vehicle will start when an authorised person places their finger on the fingerprint sensor. If the fingerprint does not match the registered fingerprint, then the vehicle does not start.

In the event of a theft, we can know the location and position of the vehicle by using GPS/GSM. The GPS collects the data and stores the data in the Arduino at determined time intervals. The user acquires the real-time location of the vehicle using the GPS/GSM module. In case of an emergency, an alert message sent to the user, as shown in the below figure. GSM is used to send the stored data to the user, and it involves location, tracking, altitude, elongation, and slack. By utilising this data, we can find out the exact location of the vehicle using Google maps. The advantage of this system is that we can control the ignition of the vehicle from our smart phone by using the blink Android app. The owner can easily decommision his vehicle as soon as he comes to know about the theft of the vehicle. When the user sends a command to Arduino through GSM, the gas pedal cannot be moved, and the flow of fuel stopped, and the vehicle cannot move. This system helps to unlock the vehicle without any keys, disables the security system, and provides adaptable protection to the vehicle.

8.5 Results and Discussion

Figure 8.8 shows vehicle status displayed for in a specified mobile phone number if anybody trying to steal and have an accident with a vehicle. This project provides us with a clear understanding of how automatic control works in motorcycles and also increases security to avoid theft of vehicles by adding anti-theft security into the system. The system is able to provide better flexibility by including an Arduino Uno model and is mostly effective.

8.6 Conclusion

This project gives examples about how security is being included in vehicles. This project provides extensions like the accelerometer and IoT for increasing the security level. This security mechanism includes the best-performing components like

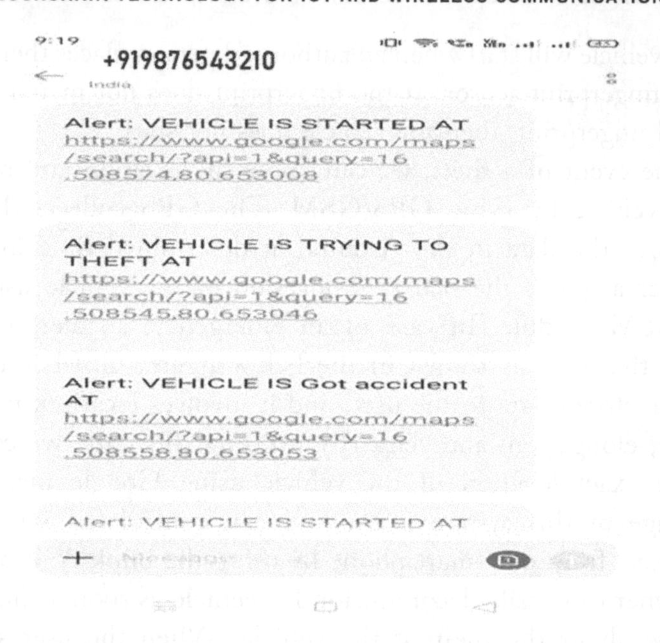

Figure 8.8 Vehicle status displayed in specified phone number.

fingerprints, GSM/GPS, etc., in the field of security. By including this project in motor vehicles, it removes the need for a key to start the vehicles. The fingerprint itself is able to start the ignition of a motor vehicle. We have specified GPS for tracking the live coordinates and GSM for alerting the user about theft via a text message. This project has given access to control the ignition of the vehicles using mobile phones.

References

[1] Dey, M., Arif, A., Mahmud, A., "Anti-theft protection of vehicle by GSM & GPS with fingerprint verification" In: International Conference on Electrical, Computer and Communication Engineering (ECCE); 2017. pp. 1–5.

[2] Kumar, D. A., Manohar, S., Hari, G. T. S., Gayatri, G., Venkateswarlu, A., "Detection of COVID-19 from X-RAY Images using Artificial Intelligence (AI)" 2022 International Conference on Intelligent Technologies (CONIT); 2022. pp. 1–5.

[3] Soaid, M. F., Kamaludin, M. A., Megat Ali, M. S. A., "Vehicle location finder using global position system & global system for mobile" In: Control & System Graduate Research Colloquium (ICSGRC); 2014. pp. 279–284.

[4] Verma, P., Bhatia, J. S., "Design and development of GPS-GSM based tracking system with google map based monitoring" *Int. J. Comput Sci. Eng. Appl. (IJCSEA)*, 2013; 3(3): 1–12.

[5] Ijjada, S. R., Sharma, A., Babu, M. S., Kumar, D. A. (2020). "SS<30mV/dec; Hybrid tunnel FET 3D analytical model for IoT applications" *Materials Today: Proceedings*. 10.1016/j.matpr.2020.09.367.

[6] Nalina, V., Sandesh, A. S., Sequiera, R. D., Jayarekha, P., "Cloud based multiple vehicle tracking and locking system" In: 2015 IEEE International Advance Computing Conference (IACC); 2015. pp. 1–6.

[7] Ajay, K. D., Rao, I. S., Murthy, P. H. S. T., "Performance analysis of 3G SOI FinFET structure with various fin heights using TCAD simulations" *Journal of Advanced Research in Dynamical and Control Systems*, 2019; 11(2): 1291–1298.

[8] Wimberly, H., Liebrock, L. M., "Using fingerprint authentication to system security; An empirical study" In: IEEE Symposium on Security and Privacy; 2011. pp. 1–9.

[9] Kumar, A. D., Ijjada, S. R., "A novel design of SOI-based fin gate TFET" 2021 2nd Global Conference for Advancement in Technology (GCAT), 3 November 2021, IEEE, Bangalore, India. 2020., 10.1109/GCAT52182.2021.9587599.

9

Soft Sensor-Based Remote Monitoring System for Industrial Environments

AJAY KUMAR DHARMIREDDY[1], P. SRINIVASULU[2], M. GREESHMA[3], AND K. SHASHIDHAR[4]

[1]*Department of ECE, Asst. Professor, Sir C.Reddy College of Engineering, Eluru, India*
[2]*Department of ECE, Asst. Professor, Sir C.Reddy College of Engineering, Eluru, India*
[3]*Department of ECE, Asst. Professor, Sir C.Reddy College of Engineering, Eluru, India*
[4]*Department of ECE, professor, Lords Institute of Engineering and Technology, Hydrabad, India*

Contents

DOI: 10.1201/9781003269991-9

9.1 Introduction

The Internet of Things (IoT) has the ability to change the world. It can generate data about linked equipment, process information, and pass judgement; in other words, the IoT outperforms the Internet. It is also commonly used in the industrial sector. Because safety and security are so important in the industrial sector, we are putting in place an IoT-based industrial surveillance system to stop dangerous gases from leaking and causing explosions by accident.

It will also serve as an access control system. This system works by sounding an alarm when there is a gas leak and showing the approved gas limit and the warning level. To detect common circumstances of the machinery within a business, it plays an important role. Whenever there are any radical changes within instrumentation, unless humans perceive chemical heating of the equipment or the average temperature of a chemical surpassing the temperature controller, and make adjustments so the heat manages to keep on target, then the chemical may overflow or incinerate. This system continuously monitors the equipment with different parameters like temperature, percentage of gases in the air, and pressure. Due to this, we are able to monitor all the time, so humans can be able to identify the changes. If designers observe any abnormal conditions, they can alert and take immediate action on the corresponding problem. It saves many lives a day. The gases released from chemicals cause health problems for those who are close by [1,2]. To overcome this type of problem, we introduced this system. This system continuously monitors the hardware with different parameters like temperature and percentage of gases.

9.2 Literature Survey

Major issues/problems in domestic life as well as industries like gas leakage, fire detection, and temperature increases are causing disasters. To avoid these disasters in advance, an alternative idea is discussed in this paper – using sensors, such as a temperature sensor to detect an abnormal rise in temperature, and a gas sensor to detect gas leakage, and a flame sensor to detect fire in the industry. The detected issues can be rectified by using some remedies such as when temperature rises more than the threshold value, an LED will turn on. When a gas leak is detected, an exhauster fan turns on automatically, when fire is detected, a buzzer will produce a high-frequency sound as an alert. In this way, industrial disasters can be prevented. The author in one paper created an IoT-based smart industry monitoring system by using Raspberry PI3 to explain the same parameters but using a Raspberry Pi board [3,4].

9.3 Methodology

This project design process is based on data, which are collected through several types of sensors like flame sensor, gas sensor, and temperature sensor (Figure 9.1). These sensors are installed in particular locations, such as where there is a possibility for gas leakage. The temperature sensor is used to track the temperature of industrial equipment like boilers and display the readings via the Thingspeak website; the flame sensor is used to detect fire. These sensors gather data continuously and communicate the data to the Node MCU [5-7].

The Wi-Fi module is programmed with a specific threshold value. If the value is less than the threshold value, the situation

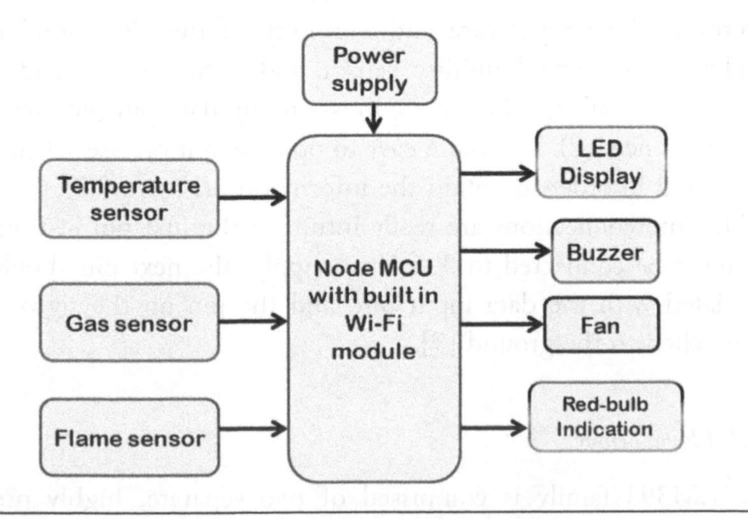

Figure 9.1 Block diagram of IoT-based industrial welfare automated system.

is normal. If it exceeds the threshold value, the Wi-Fi module sends a signal to the corresponding output. If the temperature is high, an LED will glow to indicate that there is a rise in temperature. If gas is leaking, the exhauster fan will start rotating in order to decrease the concentration of gas in the air equipment. If a fire is detected by the flame sensor, a corresponding buzzer will produce sound continuously until the fire is extinguished [8].

All the sensor data are shown 24 * 7 in the cloud-based platform. Over IoT, data are sent and shared using a Wi-Fi module. From our Wi-Fi network, we can control the Wi-Fi module. The Wi-Fi module allows microcontrollers to connect to wireless networks. Thingspeak.coma is a cloud-based web platform used to send the data from sensors like gas, flame, and temperature sensors to the cloud. The Thingspeak platform uses data analytics and data visualization tools to show the graphs corresponding to those sensed parameters. By using this website, we can monitor the industrial parameters anywhere in the world. We have designed a fully autonomous IoT wireless sensor-actuator network for monitoring the industry environment.

9.3.1 Temperature Sensor

The DHT22 is a straightforward digital thermometer and temperature and humidity sensor, and it doesn't cost a fortune. In order to determine the temperature and humidity of the air around it, it employs a sensitive humidity sensor and a thermistor, and then it emits a digital signal on the corresponding data pin (no analogue input pins needed). It is quite easy to operate, but precise scheduling is required in order to obtain the information (Figure 9.2).

The interconnections are really intuitive: the first pin just on the left must be connected to 3–5 V of supply; the next pin should be associated with the data input pin, and the pin on the right must be attached to the ground [9].

9.3.2 Flame Sensor

The LM393 family is comprised of two separate, highly precise voltage comparators that are able to work with either a unified or

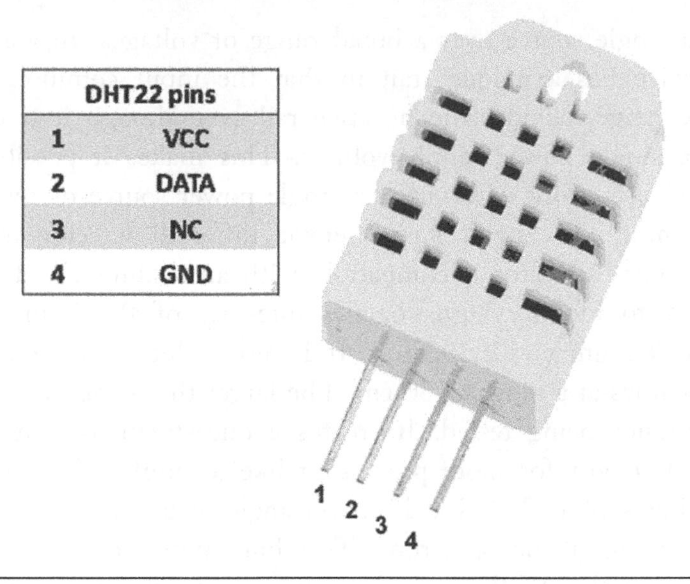

DHT22 pins
1 VCC
2 DATA
3 NC
4 GND
1
2
3
4

Figure 9.2 Temperature sensor.

Figure 9.3 Flame sensor.

divided source. Such circuits are intended to provide a common mode of range-to-ground level functioning while using a common power source (Figure 9.3).

This system is made up of two different low-power voltage comparators that are completely independent of one another. They were developed in such a way that they could function from a single source over a broad range of voltages. In addition, the device has a unique trait in that the input common-mode voltage range includes the negative rail even though it is driven from a single power supply voltage. This makes it possible for the device to work with either a single power source or two separate ones. This small flame sensor infrared detector module flame source detector is compatible with an Arduino, and it can be used to identify burns or the intensity of the illumination within 760 nm and 1,100 nm. It is also helpful for identifying lesser flames at a range of 80 cm. The larger the flame, the farther the distance being tested. It creates a one-stream output signal at the D0 port for more processing like a burglar alarm or any switching system. It has a detection angle of 60 and is very sensitive to the flame spectrum. The blue potentiometer that is located on the PCB allows for the sensitivity to be adjusted as needed [10].

9.3.2.1 Gas Sensor (MQ6) The MQ-6 LPG, hydrocarbons, and acetylene sensor is a photonic detector that can recognise the existence of LPG, hydrocarbons, and flammable gases at intensities that vary from 100 ppm to 10,000 ppm, which is suitable for recognising leakage (Figure 9.4).

The MQ-6 LPG-Isobutane-Propane gas sensor detects the concentration of gas in the air and displays its measurement

Figure 9.4 Gas sensor.

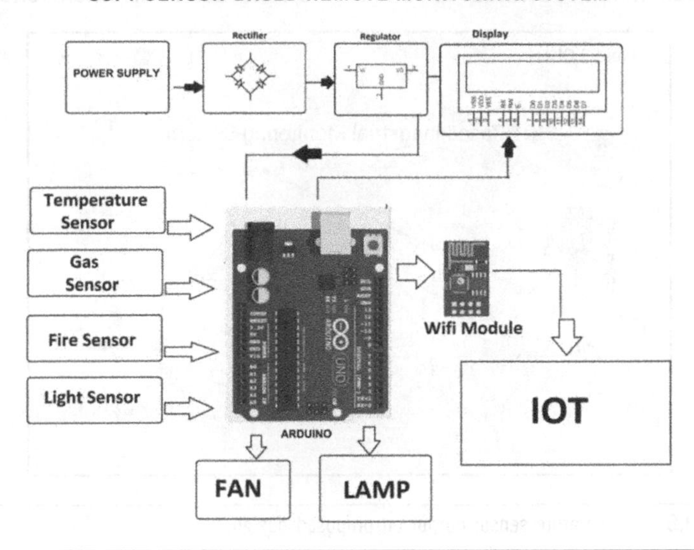

Figure 9.5 Schematic diagram for industrial welfare automation system.

as an analogue voltage. The sensor's simple analogue voltage interface needs just one analogue input pin from your micro-controller. The sensor has a temperature range of -10°C–50°C, and its current draw at 5 V is less than 150 milliamperes (Figure 9.5).

9.4 Results and Discussion

The collected sensing data from sensors are stored in a CSV (comma-separated values) file on a web server. The Thingspeak web service is employed to evaluate the proposed system's IoT components. This server is frequently updated within 20 seconds. The server takes data in the form of fields, which are visually displayed using data visualisation and data analytics, as seen in the figures below. You can also delete and remove any channel from this web server if you need to. This lets you handle huge amounts of information (Figures 9.6–9.8).

9.4.1 Temperature Sensor Output

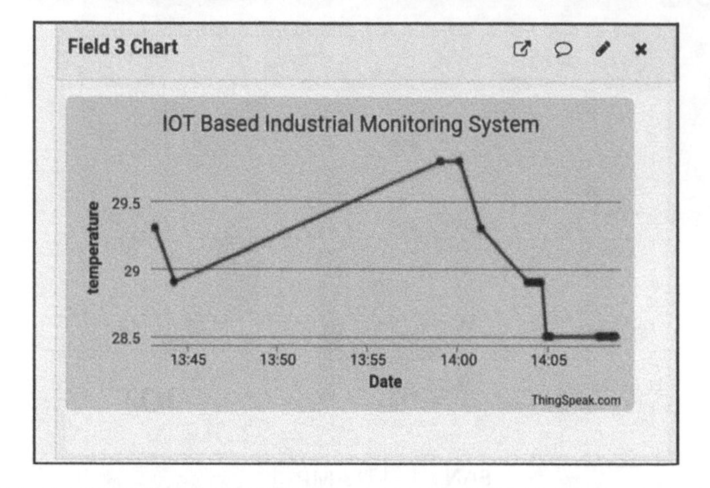

Figure 9.6 Temperature sensor output on proposed design.

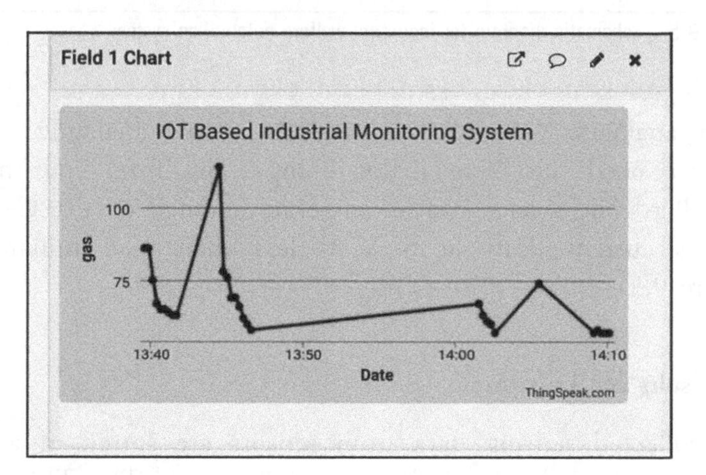

Figure 9.7 Gas sensor output in proposed design.

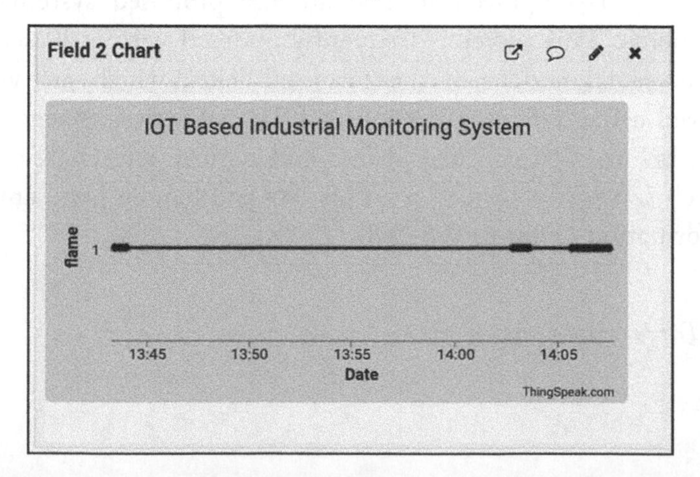

Figure 9.8 Flame sensor output in proposed design.

9.5 Conclusion

By working on this project, we aim to gain in-depth knowledge from scratch in the technical aspects of the IoT and digital electronic systems. The core objective of our industrial monitoring system is achieved with digital integration of sensors with optimum value precision and micro-controllers. We have incorporated sensors for detection of environmental parameters when they reach undesired conditions. The observed data is communicated to the web servers. Node MCU is used for processing and allocation of the data that is collected from various sensors. Therefore, we have included an MQ-6 gas sensor, an LM2903 flame sensor, and DHT22 temperature sensors for receiving data. We have also used a cloud-based Thingspeak website for the collection of data. The importance for preventing undesired hazards and activities in management's absence is the key novelty of our project, and we believe it would help industries reduce these accident rates effectively if it deployed. With ongoing industrial modernization these days, the demand for digital and automation systems to perform functions, especially in the information technology and algorithmic sectors, has risen exponentially, and the application of IoT in their operational mechanisms paves the way for development and feasible innovation.

References

[1] P. Gopi Krishna, K. Srinivasa Ravi, P. Hareesh, D. Ajay Kumar, H. Sudhakar, "Implementation of bi-directional blue-fi gateway in IoT environment" *International Journal of Engineering & Technology*, 7(2), pp. 97–102, 2018.

[2] D. A. Kumar, S. Manohar, G. T. S. Hari, G. Gayatri, A. Venkateswarlu, "Detection of COVID-19 from X-RAY Images Using Artificial Intelligence (AI)" 2022 International Conference on Intelligent Technologies (CONIT), pp. 1–5, 2022.

[3] M. Kavitha, S. H. Raju, S. F. Waris, A. Koulagaji, "Smart Gas Monitoring System for Home and Industries" International Conference on Recent Advancements in Engineering and Management (ICRAEM-2020), Vol. 981, pp. 9–10, October 2020.

[4] P. H. S. Tejomurthy, A. Kumar, P. V. T. Lokesh Kumar, "Design of a NEMS cantilever sensor for explosive detection" *International Journal of Engineering and Advanced Technology*, 8(6S2), pp. 912–, 2018.

[5] S, S. R. Ijjada, A. Sharma, M. S. Babu, D. A. Kumar , "SS<30mV/dec; Hybrid tunnel FET 3D analytical model for IoT applications" *Materials Today: Proceedings*. 2020. 10.1016/j.matpr.2020.09.367.

[6] P. Sharma, S. Prakash, Real Time Weather Monitoring System Using Iot *In ITM Web of Conferences* (Vol. 40, p. 01006), EDP Sciences, 2021.

[7] K. D. Ajay, I. S. Rao, P. H. S. T. Murthy, "Performance analysis of 3G SOI FinFET structure with various fin heights using TCAD simulations" *Journal of Advanced Research in Dynamical and Control Systems*, 11(2), pp. 1291–1298, 2019.

[8] B. Tamilarasi, P. Saravanakumar, "Smart sensor interface for environmental monitoring in IoT" *International Journal of Advanced Research in Electronics and Communication Engineering (IJARECE)*, 5(2), pp. 1–9, February 2016.

[9] A. D. Kumar, S. R. Ijjada, "A Novel Design of SOI-Based Fin Gate TFET" 2021 2nd Global Conference for Advancement in Technology (GCAT), 3 November 2021, IEEE, Bangalore, India. 2020. 10.1109/GCAT52182.2021.9587599.

[10] Ms. S. C. Padwal, Prof. M. Kumar, "Application of WSN for Environment Monitoring in IoT Applications" International Conference On Emerging Trends in Engineering and Management Research (ICETEMR-16) – 23 March 2016.

10

The Impact of the Internet of Things on Measurement, Monitoring of Power System Parameters in an LFC-DR Model

P. SRIVIDYA DEVI AND PRAVEEN JUGGE

Gokaraju Rangaraju Institute of Engineering and Technology, Hyderabad, India

Contents

10.1 Introduction

Load frequency control (LFC) is an important concern in power system operation control for delivering adequate and consistent electric power with standard quality. In an IEEE PES committee report [1], the authors suggested scientific models that provide adequate representation for thermal, hydro, and fossil-fired generations. In most stability studies, which are mathematical studies for the turbine systems, speed-governing models are performed. Kothari and Nanda [2] discussed only discrete-mode for a two-area reheat thermal system for the LFC model with area control error (ACE) in depth, but only with

DOI: 10.1201/9781003269991-10

conventional controllers. Issues related to the real-time power system control and also constrained feedback control schemes are presented by R.R. Shoults, J.A. Jativa [3], and M. Aldeen [4]. Many authors and various studies, such as Kundur P. [5], gave the overall picture about the stability and control in power systems. Julio Concha and Aldo Cipriano [6] proposed a design method for stable operation using fuzzy and LQR controllers. For the past two decades, robust analysis and design has been carried on the LFC and regulation [7]. Ray G. et al. [8] and Al-Hamouz et al. [9] analysed new approaches for load frequency controllers of an interconnected system. Many authors like M. Azzam et al. [10], Ghoshal, S. P. [11], and M. Hovd et al. [12] presented in a well-elaborated way automatic generation control (AGC) for a tie-line power flow frequency control for a multi-area power system (MAPS) with a controller's vital increase planning.

From the past few years, the controllers became advanced, intelligent, and adaptive like fuzzy logic (FL), where they play a vital role. E. H. Mamdani et al. [13–15] proposed mamdani fuzzy rules by models, and Takagi-Sugeno [16]-based fuzzy relational models. It is known that conventional controllers like proportional integral derivative (PID) controller are less superior than fuzzy logic controller (FLC) as they exhibit better performance in dynamic mode. Ilhan Kocaarslan et al. [17] and B. Anand et al. [18] designed FL algorithms for controlling different plants and their applications of FL controller systems. Nowadays, many authors state that the Type-2 Fuzzy Logic Control (T2FLC) is more advanced intelligent control. K.R. Sudha et al. [19–21] designed a decentralized LFC model power system with generation rate constraint (GRC) for MAPS using Type-2 fuzzy approach including SMES units. H. Bevrani et al. [22,23] have given the overall view on decentralized AGC with robust control in a power system that is restructured.

It is widely known from the literature that balancing generation and demand results in frequency regulation in the power system. Spinning and non-spinning reserves are used for this. High adoption of variable renewable energy sources is anticipated for the future smart grid. Demand response (DR) and distribution grid operations provide potential and problems, according to Medina et al. [24,25]. In a pilot study combining smart meters and remote load control, price was based on the hourly spot price paired with a time-of-day network fee, and a token given to the users to indicate peak hours, shared by H. Saele et al. [26]. To help maintain a balance between generation and load, A. Brooks et al. [27] provided the demand dispatch employing real-time control of demand. The earlier writers had a few drawbacks, which S. A. Pourmousavi et al. [28–34] have highlighted. In-depth presentations have been made on modelling DR in retail power markets and frequency control of microgrids by DR. This leads to the presentation of a real-time DR for primary regulation. DR has advantages and disadvantages, and LFC-DR is introduced.

But with the increasing demand of the renewable energy resources there exists a rise of distributed generations (DGs) in each area. In this chapter, thus, balancing of demand and supply of the power generations is achieved by introducing the DR control loop to the traditional LFC for a power system. This facilitates LFC-DR. The DR provides ancillary services (AS) for regulatory reserve. An interconnected power system is a combination of different control areas, in which each area relates to a transmission line called tie-line. Such power systems are called interconnected power systems. In each control area, all generators are assumed to form a coherent group.

The power system is subjected not only to the local variations of random magnitude but also to communication delay latencies (CDL) with the DR intrusion. The IoT has attracted much attention recently and has pained a beautiful picture for future smart life. Qian Zhu et al. [35] gave a brief overview on IoT Gateway and bridging wireless sensor network into IoT. L. Atzori et al. [36] addressed different visions on the IoT, its aiding factor for this promising growth of paradigm. The various parameters are integrated with technologies to provide a solution for the future next generation with the Internet.

Wen-Long Chin et al. and several other authors [37–39] developed many smart appliances, sensors, and actuators with two-way communications. However, one of the major drawbacks in integration is security in adopting the vision of a smart grid. Information on parameters for the power monitoring and utilization are widely considered in a smart grid. The smart grid DR can reorganize the user's energy consumption, which results in reduction of operating expenses. The DR is useful in addressing the future challenges. By smart direct load control, power outages can be minimized in the grid, thus focusing on the problems and challenges related to DR control in the smart grid. The DR frequency control uses the IoT which provides real-time load control. Shedding of loads and controlling the loads can be directly done in a smart grid with different learning algorithms, which have presented the usage of demand response. There arises several research questions:

- What is the necessity of the DR?
- Why only use the intelligent controllers?
- What is the importance of the IoT?

There are a number of surveys on the performance of distributed and cloud-based approaches on optimal control, thereby achieving price-based DR schemes. In present time, smart utilization of energy resources for emerging smart cities is available. Thus, a high-priority load is connected to maintain the frequency regulation within power threshold limits. The IoT concept nowadays enables devices to connect over the Internet with each other. These all are connected, controlled, and remotely analysed. Electricity demand is constantly increasing for various reasons like use by hospitals, industries, agriculture, households, and many more. Handling this requirement became complex as a part of electricity reliability and maintenance. So, DR integrated with the local load management and measurement is important.

10.2 Role of Demand Response with IoT for Future Smart Grid

As the population's interest in power from younger generations grows, so does the need for innovation enhancement. The suggested framework uses IoT technology to give conventional energy meters a specific use. Similarly, there are several concerns that address, for example, power theft and meter altering, which cause financial hardship for the country. Checking these crimes, enhancing power use, and decreasing of wastage of power are the significant goals that lie ahead for a superior framework.

The current framework relies upon more human contribution for different cycles than should be done in power system operation and control. This is a tedious cycle. To address every one of the referenced requirements, there is surely a fostered framework in view of IoT innovation [40,41].

For example, considering smart meters at the utility level using the Internet, there are mainly concerns on different prospective and objectives:

1. To give smart meter perusing a quick premise.
2. To involve and use the power in a streamlined and optimized way.
3. To lessen the wastage of power.

Similarly, it is useful and classified based on the service providers in two ways:

1. Service end
2. User end

The information from the framework is shown on a web page that can be reached by both the customer and service provider for further monitoring and control of parameters in the power system.

The world is moving toward a more advanced society where end users have access to local generations that are powered by renewable resources and cutting-edge information technology. As a result, there is currently a transition taking place between the conventional technique of power generation and distribution, where end users of electricity have usually been passive in their interaction with energy markets, and a new strategy that incorporates their active participation. This innovative approach makes use of DR, distributed energy resources (DER), and renewable-based power, all of which are quickly being embraced by end users where incentives are high.

There are more DERs based on renewable energy in the power system on the demand side of the grid [42]. DR is one of the resources that all renewable technology resources, including wind energy, electric vehicles, and solar photovoltaic systems, fall under [43]. Integrating new technology resources into the current infrastructure and energy markets is one of the biggest issues facing power systems operators globally. They lack the proper controls and monitoring systems for the low-voltage networks to which these resources are connected. Due to the intermittent nature of renewable DERs, the issue can occasionally get worse. As a result, sometimes it is difficult to notice changes in frequency and voltage [44]. Thus, there is a chance that power quality and system stability could be compromised. To tackle the issues regarding high penetration of DERs, one promising approach is engaging the customer in the operation of these technologies.

However, most end users in residential areas no longer have regular access to dynamic charge alerts of power, which limits their ability to participate in electricity markets. Having said that, the adoption of DR programmes has made it easier for business and commercial end users to establish a presence in marketplaces [45].

To maximise benefits for them and other players, like utilities, corporate and commercial clients must coordinate. The efficiency of integrating distributed DR and customer flexibilities as a resourceful capability for reducing customers' overall power costs [42,46,47] or improving grid operation [48–51] and funding on electricity transport system is also demonstrated by certain study. A DR aggregator makes it possible to regulate those goals for this purpose.

Grid operators deploy DR systems to keep the system affordable and stable during times of high demand, high DER generation, or high-power cost score [52]. These plans take advantage of the ability of the end user to react to operator signals by transferring or lowering certain transmissions or generations in exchange for a benefit or incentive [52]. The advantages of managing these loads or generations include lower bills and rewards for the end user, stable market volatility, the preservation of network infrastructure, the effect of energy, improved network reliability and stability while lowering marginal costs during peak events, and the flexibility of systems that can be used to integrate renewable energy technologies. Traditional energy providers may be effectively ousted from wholesale energy markets by demand-side bidding through DR because the running costs to enable DR are much lower than the operating costs of a single unit.

A smart meter is a device that allows to communicate information to the end user of electricity, such as device energy consumption measurements, load configuration, power measurement, frequency, tariff schedule. according to usage time, trip events, voltage levels, phase loss and asymmetry. With this additional information, users may respond to electrical signals and make wiser choices regarding their energy use, becoming active players in the electric market.

While low-voltage networks are typically not monitored by traditional power systems, smart meters offer a monitoring mechanism that makes them visible and traceable, which is crucial for successfully integrating DER integration. While the deployment of this technology is a key factor in population engagement in R&D, the reality remains that these users are hardly motivated for the reasons outlined above. As a result, activating the full potential of people requires third parties to develop tailor-made products that allow users to contribute to decision-making without too much difficulty implementing offload guidance.

Regulation of frequency is the primary objective of the power grid management, which balances the demand and supply. The integration of DR and traditional frequency regulation results in the distributed DR frequency control, which is known as LFC-DR model. The use of DR control further enhances the load participation. The smart intelligent controllers give superior performance than classical controllers in settling the disturbances in the system.

The information of the power monitoring and utilization are widely considered in a smart grid. The smart grid DR can reorganize the user's energy consumption, which results in reduction in operating expenses. Thus, DR is useful in addressing the future challenges. By smart direct load control, power outages can be minimized in the grid (Figure 10.1).

Even during the contingencies, the power system dynamics operate in synchronous regime where the whole system stays as a distinct system with wide frequency. There are models in a power system with an aggregate from the generation of power and power demand.

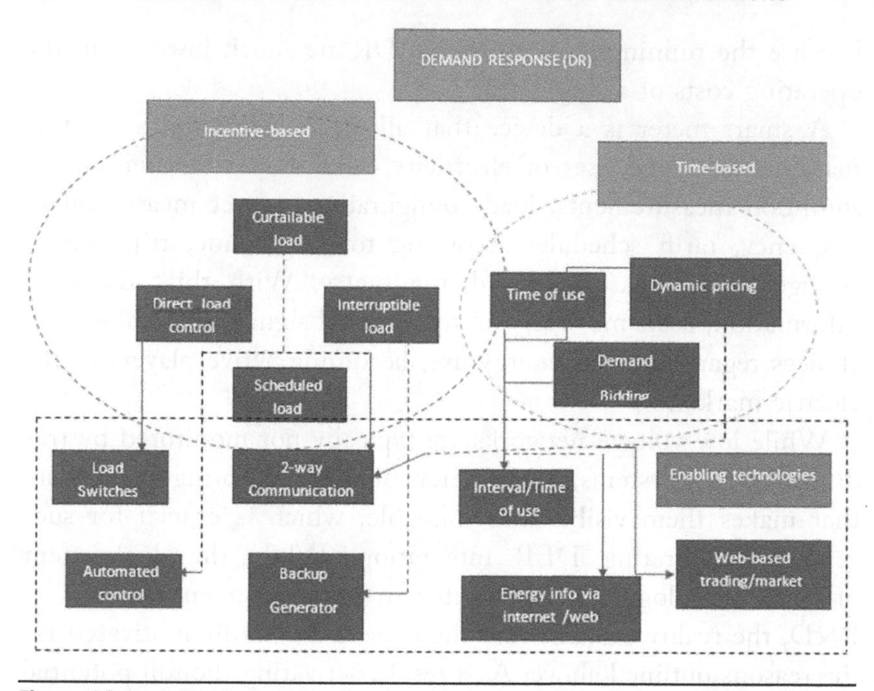

Figure 10.1 Demand response categories with the enabling technologies [107].

The demand power is modelled as a function of frequency. Let *f(t)* denotes the system frequency with nominal value of f_O at time $t = 0$. From the previous model, it is well known that deviations in frequency from their nominal values are realistically small. The consumption of power is aggregated and expressed in terms of system frequency as a linear function:

$$P_d(t) = P_0 + \left(\frac{f(t) - f_0}{f(t)} \right) K_f P_0 \qquad (10.1)$$

where P_d = aggregated demand; K_f = frequency damping coefficient; P_O = represents consumption of power aggregated under the nominal values of frequency and voltage at time $t = 0$. The DR can only be used to recover the system frequency, and generation side frequency controllers such governors and automatic generation controllers are presumptively disabled. As a result, the power dynamics are the only factor affecting frequency dynamics. The IoT is used for real-time load control as part of the DR frequency control. Thus, this chapter focuses on the problems and challenges, related to DR control in a smart grid. A prototype was developed for monitoring the power system parameters with ESP8266 and Thingspeak.

10.3 Performance Analysis and Curves of LFC-DR Model

Various case studies were performed on the LFC-DR model of the interconnected area power system to verify the robustness of the proposed method. To perform simulation studies, based on discussion as in [34], since DR is available in the LFC, more reliable frequency tuning can be achieved, since the control loop of the DR can be supplemented as another additional control loop presented in [53]. In the absence of supplementary control of LFC, frequency tuning performance can be guaranteed per DR loop, if the availability of DR resources is sufficient. The studies are carried out in detail in [53,54]. The performance curves under steady state with the control effort "α" for the changes in frequency deviation with DR and without DR are demonstrated [54]. In the simulation, 0.1 p.u. disturbance of load (10% load perturbation) and with control effort α = 0.8 (which says DR participation is 20%) was applied to the power system of multi-area with a communication delay latency of Td = 0.1. The proposed method shows better performance, and the comparative analysis of the curves are shown in Figures 10.2 and 10.3.

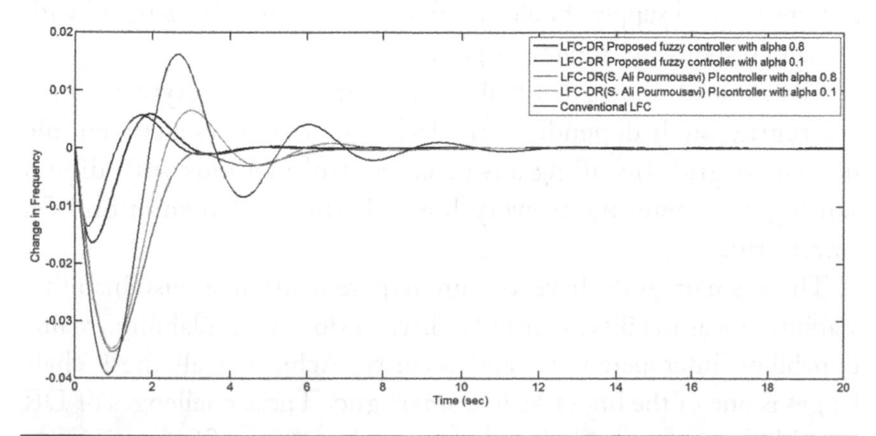

Figure 10.2 Frequency deviation for conventional LFC and LFC-DR models (with PI and proposed fuzzy logic controller).

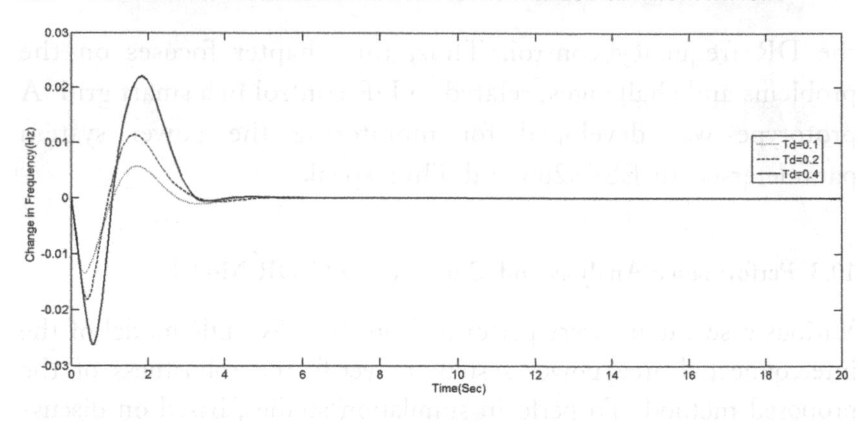

Figure 10.3 Frequency deviation for LFC-DR model with fuzzy logic controller for different delay latencies.

The next part implements the hardware prototype of direct load control and monitoring with measurement over the cloud for future grid analysis and prediction.

10.4 Hardware Implementation Prototype of Direct Load Control through IoT

Load control in DR became more prolific and is one the challenging task for the smart grid, with the progress of communication infrastructure. Real-time communication between users and the utility operator is now two ways after various taxonomy methods in control of demand. The focus is on two aspects: grid operational cost should be minimized, and supply should not fluctuate meeting the demand with demand response, in serving at peak load times.

Figure 10.4 shows control in a smart grid. Everything in a smart grid purely depends on the IoT. The IoT plays a crucial role in a smart grid. Installing advanced control methods with digital sensing and metering at every house is the implementation of a smart grid.

These smart grids have certain requirements like sustainability, stability, measurability, controllability, flexibility, availability, maintainability, interoperability, and security. Achieving all these challenges is one of the big tasks in a smart grid. These challenges of DR are well focused, considering the power demand profile to minimize the operational cost.

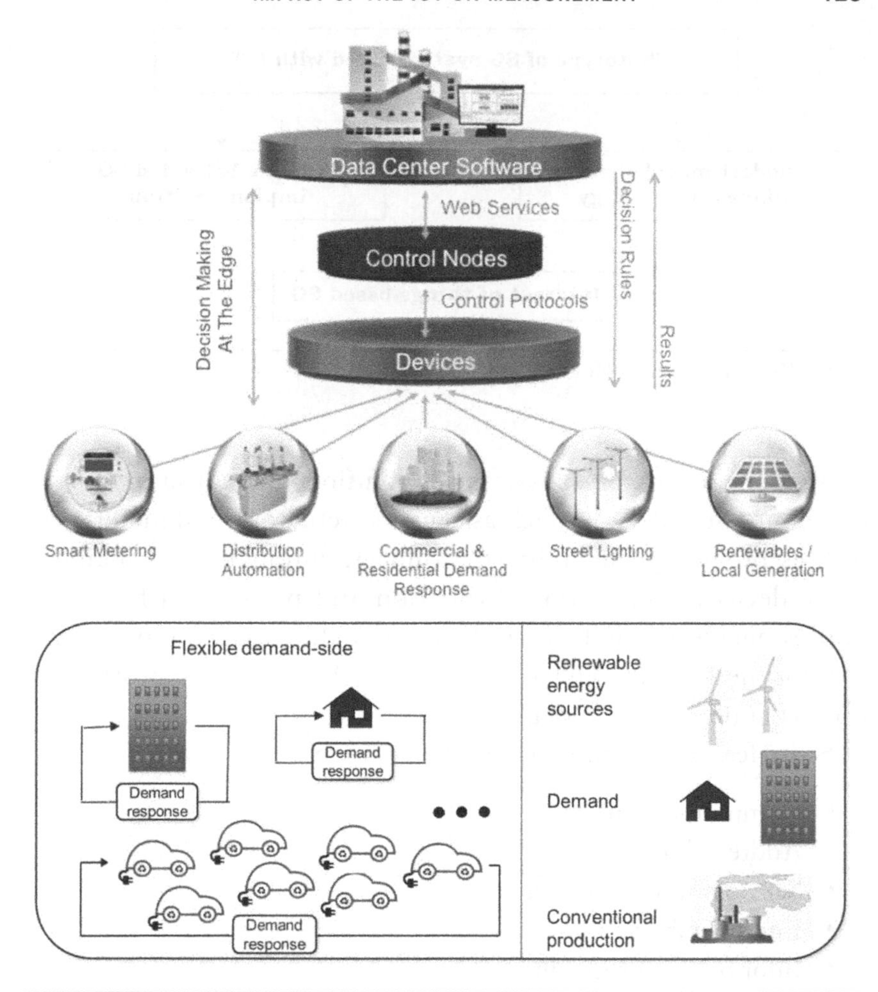

Figure 10.4 Demand response smart grid control network.

To address issues with unidirectional information transfer, energy waste, rising energy demand, reliability, and security, the conventional power system is being changed into a smart grid (SG). Through two-way communication between service providers and customers, SGs facilitate the flow of energy in systems for the production, delivery, and use of electricity. SGs use a variety of network monitoring, analysis, and control tools that are widely used in power plants, distribution hubs, and user facilities. In order to link, automate, and monitor these devices, an SG is necessary. The IoT is used to do this. IoT helps SG systems support various network functions throughout

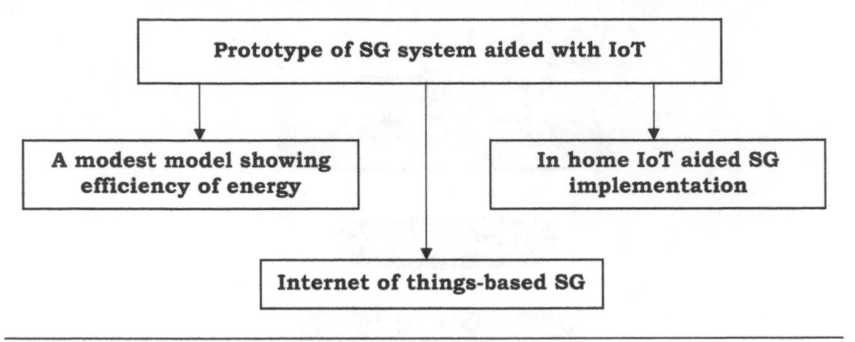

Figure 10.5 Prototypes of SG (smart grid) system.

power generation, transmission, distribution, and consumption by combining IoT devices (such as sensors, actuators, and meters), as well as providing the ability to connect, automate, and monitor those devices. Thus, two information and power transfer capabilities enable the IoT to control smart devices. This paves the way for the IoT enabling a DR. Figure 10.5 shows the prototypes for IoT aides in SG systems.

Some features required in smart devices are:

- Communication
- Addressability
- Sensing and Actuation
- Interoperability
- Information Processing
- User Interfacing

To address the challenges, a prototype board was developed with the At-mega 328 microcontrollers (Arduino UNO /Mega) with ESP8266 Wi-Fi module or NodeMCU that has inbuilt Wi-Fi etc. Using the higher-level MODBUS implementation for data logging with RS485 interface for smart meter connectivity, more than 40 power system parameters can be read. As it is, with At-mega 328 microcontroller the programming is done with open-source software. The block diagram for the implementation of the smart metering and monitoring for power system parameters in the cloud via esp8266 is shown in Figure 10.6. The overall hardware implementation using the voltage and current sensors is shown in Figure 10.7.

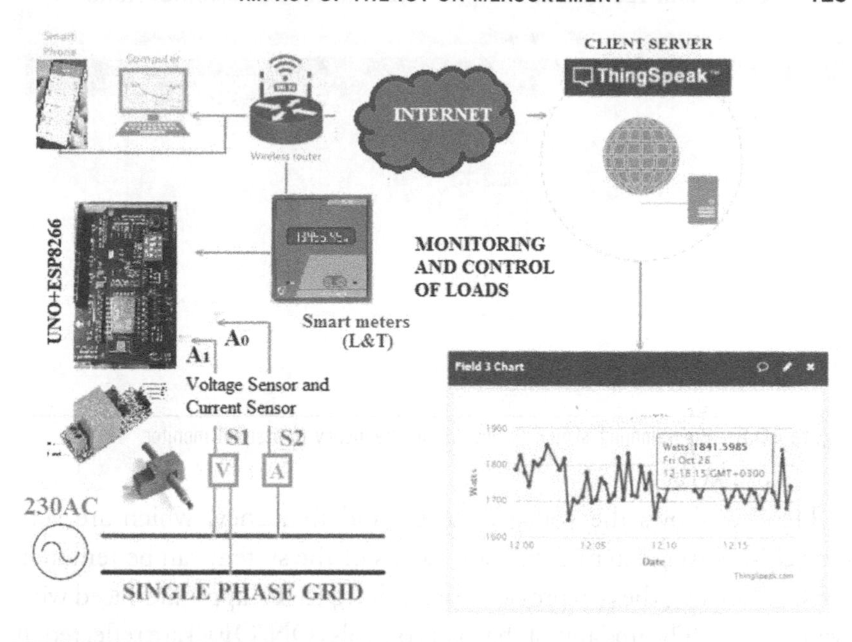

Figure 10.6 Block diagram for the hardware implementation for IoT.

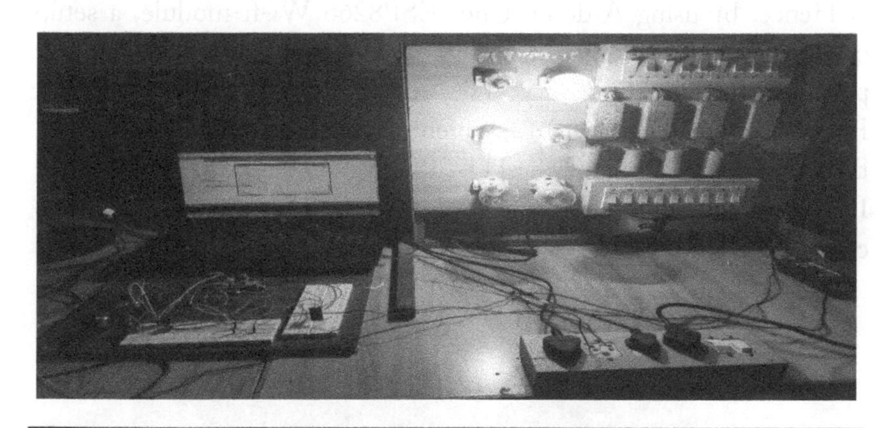

Figure 10.7 A prototype showing the hardware implementation of the circuit.

The output of frequency fluctuations (as one of the parameters) is shown in Figure 10.8 in a serial monitor. The parameters like voltage, power, power factor, and frequency, etc. are pushed on to the cloud using the web services and mobile app developed by MIT AI2 Companion. The continuous monitoring of these values graphically can be seen through API channels in Thinkspeak, interfaced with the Wi-Fi module, thus allowing the user to supervise all the parameters related to the power system.

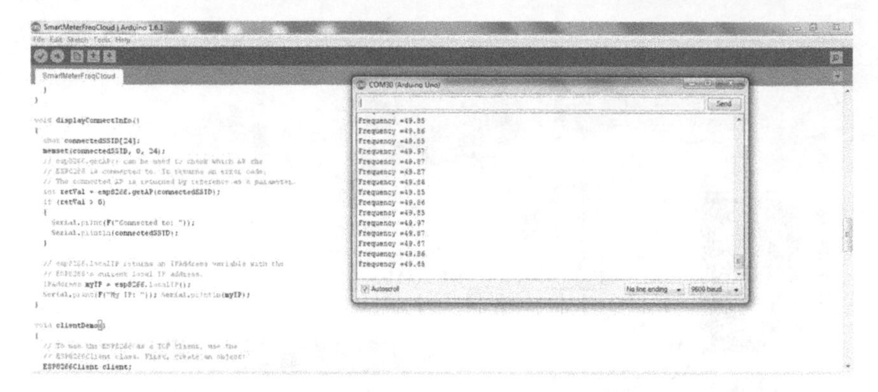

Figure 10.8 Programming sketch for monitoring frequency in a serial monitor.

Here, it shows the voltage, power, and frequency, which are very crucial in power system dynamics. Loads of the system can be remotely controlled using the app provided in Android (MIT app) interfaced with web services. The updates of the load controls (ON/OFF) are reflected in the cloud service, which is connected to the client server (Thinkspeak).

Hence, by using Arduino Uno, ESP8266 Wi-fi module, a setup of smart meter helps us to measure various values of voltage, power, power factor, and frequency. The output waveforms are shown in Figure 10.9 and Figure 10.12, which are monitored via Thinkspeak, the cloud platform, using the HTTP and MQTT protocols over the Internet or over a local area network; it is an open-source IoT application and API that stores and retrieves data from objects.

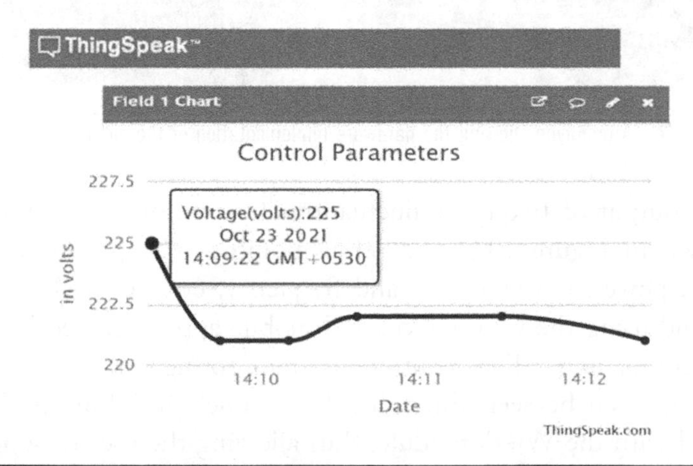

Figure 10.9 Monitoring voltage values on the cloud.

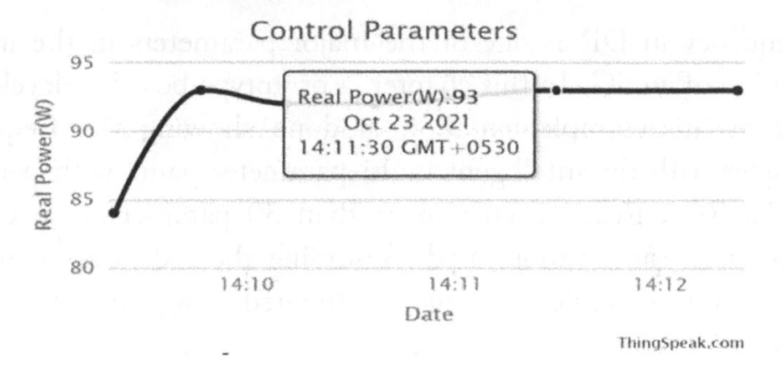

Figure 10.10 Monitoring real power values on the cloud.

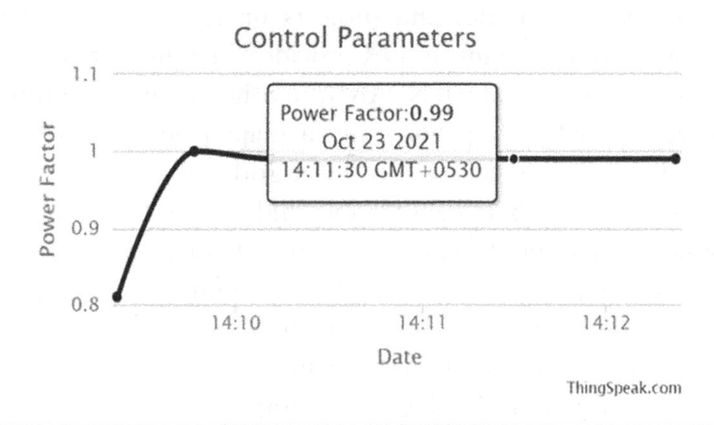

Figure 10.11 Monitoring values of power factor on the cloud.

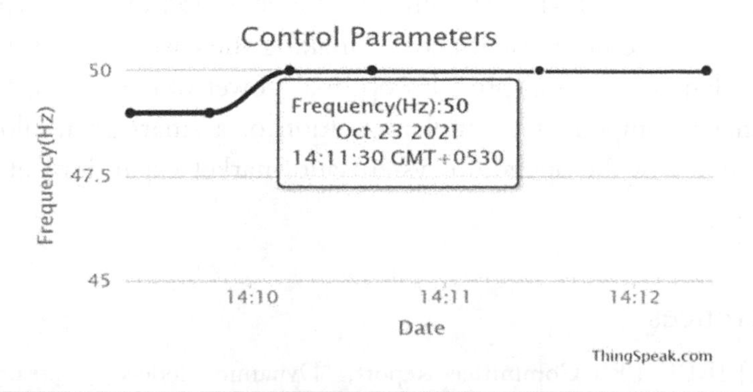

Figure 10.12 Monitoring values of frequency on the cloud.

Frequency in DR is one of the major parameters in the implementation of an SG. In this chapter, a prototype board is developed, and a real-time implementation is done showing the frequency parameter with the intelligent multi-parameter panel with parameters like Real, Reactive etc.; more than 30 parameters on energy consumption can be monitored. Assuming these data are collected for years, future predictions can be estimated using various machine learning algorithms.

10.5 Conclusions

From various case studies and projects on research, it is evident that the advanced programs in DR provide some innovative enabling technologies like smart meters, AMI, and home energy controllers, which are required for implementing such appliances and supporting coordination of automated DR in smart grid with the integration of storage devices, distributed generation, and on-site RES. Due to the lack of knowledge about the most recent technology and procedures, this results in both increased flexibility and complexity. Measurement and computational process advancements in the field of integrated electronic circuits, optimization and control systems, and information and communication technologies are important study topics that need more investigation.

With reduction of conventional power, integrating these sources with non-conventional ones and synchronizing them under the same grid is one of the major problems; implementation of these types of smart meters to domestic users and making the customers awareness of the dynamic pricing provides optimal power usage. Thus, DR is one major component in implementation of a smart grid, allowing the benefits of the operation system and market expansions at high efficiency.

References

[1] IEEE PES Committee Report, "Dynamic models for steam and hydro turbines in power system studies", *IEEE Transaction on Power Apparatus and Systems*, Vol. PAS-92, No. 6, pp. 1904–1915, November 1973.

[2] M. L. Kothari, J. Nanda, D. P. Kothari, and D. Das, "Discrete-mode automatic generation control of a two-area reheat thermal system with new area control error", *IEEE Transactions on Power Systems*, Vol. 4, No. 2, pp. 730–738, 1989.
[3] R. R. Shoults, and J. A. Jativa, "Real time power system control: Issues related to variable nonlinear tie-line frequency bias for load frequency control", In C. T. Leondes (ed.), *"Analysis and Control System Techniques for Electric Power Systems"*, Vol. 3, USA: Academic Press, pp. 377–405, 1991.
[4] M. Aldeen, and H. Trinh, "Load-frequency control of interconnected power systems via constrained feedback control schemes", *Computer and Electrical Engineering*, Vol. 20, p. 71, 1994.
[5] P. Kundur, *Power System Stability and Control.* New York: McGraw-Hill; 1994.
[6] D. P. Kothari, and I. J. Nagrath, *Modern Power System Analysis.* 4th ed. New Delhi: Tata McGraw-Hill, 2011.
[7] J. Concha, and A. Cipriano, "A design method for stable fuzzy LQR controllers", IEEE Conference, 1997.
[8] A. M. Stanković, G. Tadmor and T. A. Sakharuk "On robust control analysis and design for load frequency regulation", *IEEE Transactions on Power Systems*, Vol. 13, No. 2, pp. 449–459, May 1998.
[9] G. Ray, A. N. Prasad, and G. D. Prasad, "A new approach to the design of robust load- frequency controller for large scale power systems", *Electric Power System Research*, Vol. 51, pp. 13–22, 1999.
[10] M. Azzam, "Robust automatic generation control", *Energy Conversion & Management*, Vol. 40, pp. 1413–1421, 1999.
[11] S. P. Ghoshal, "Multi-area frequency and tie-line power flow control with fuzzy logic based integral gain scheduling", *IE (I) Journal*, Vol. 84, pp. 135–141, 2003.
[12] M. Hovd, and S. Skogestad, "Sequential design of decentralised controllers", *Automatica*, Vol. 30, pp. 1610–1607, 1994.
[13] E. H. Mamdani, "Application of fuzzy algorithms for control of simple dy-namic plant", *Proc. Inst. Elec. Eng.*, Vol. 121, no. 12, pp. 1585–1588, 1974.
[14] E. H. Mamdani, and S. Assilian, "An experiment in linguistic synthesis with a fuzzy logic controller", *Int. J. Man-machine Studies*, Vol. 7, pp. 1–13, 1975.
[15] P. J. King, and E. H. Mamdani, "The application of fuzzy controller systems to industrial processes", *Automatica*, Vol. 13, pp. 235–242, 1977.
[16] T. Terano, K. Asai, and M. Sugeno, *Fuzzy Systems Theory and Its Applications*. USA: Academic Press, 1987.
[17] I. Kocaarslan, and E. Cam, "Fuzzy logic controller in interconnected electrical systems for load-frequency control", *Electric Power Systems Research*, Vol. 27, pp. 542–549, 2005.

[18] B. Anand, and A. Ebenezer Jeyakumar, "Automatic generation control with fuzzy logic controller considering generation rate constraint and boiler dynamics", Proceedings of the International Conference on Digital Factory, (ICDF 2008), India, pp. 1539–1545, August 2008.

[19] N. Ndubisi Samuel, "An intelligent fuzzy logic controller applied to multi-area load frequency control", *American Journal of Scientific and Industrial Research*, Vol. 1, No. 2, pp. 220–226, 2010.

[20] K. R. Sudha, and R. Vijaya Santhi, "Robust decentralized load frequency control of interconnected power system with generation rate constraint using Type-2 fuzzy approach", *International Journal of Electrical Power and Energy Systems, Elsiever Publications*, Vol. 33, pp. 699–707, 2011.

[21] K. R. Sudha, and R. Vijaya Santhi, "Load Frequency Control of an Interconnected Reheat Thermal system using Type-2 fuzzy system including SMES units", *International Journal of Electrical Power and Energy Systems, Elsiever Publications*, Vol. 43, pp. 1383–1392, 2012.

[22] H. Bevrani, Y. Mitani, and K. Tsuji, "Robust decentralized AGC in a restructured power system", *Energy Conversion & Management*, Vol. 45, pp. 2297–2312, 2004.

[23] H. Bevrani, "Power System Control: An Overview" In: *Robust Power System Frequency Control. Power Electronics and Power Systems.* Springer: Springer, 2014.10.1007/978-3-319-07278-4_1

[24] J. Medina, N. Muller, and I. Roytelman, "Demand response and distribution grid operations: Opportunities and challenges," *IEEE Trans. Smart Grid*, Vol. 1, No. 2, pp. 193–198, September 2010.

[25] S. Han, Z. Xu, B. Sun, and L. He, "Dynamic characteristic analysis of power system interarea oscillations using HHT", *Int. J. Electr. Power Energy Syst.*, Vol. 32, No. 10, pp. 1085–1090, 2010.

[26] H. Saele, and O. S. Grande, "Demand response from household customers: Experiences from a pilot study in Norway," *IEEE Trans. SmartGrid*, Vol. 2, No. 1, pp. 102–109, March 2011.

[27] A. Brooks, E. Lu, D. Reicher, C. Spirakis, and B. Weihl, "Demand dispatch: Using real-time control of demand to help balance generation and load", *IEEE Power Energy Mag.*, Vol. 8, No. 3, pp. 20–29, May/June 2010.

[28] S. A. Pourmousavi, M. H. Nehrir, and C. Sastry, "Providing ancillary services through demand response with minimum load manipulation", In Proc. 43rd North Amer. Power Symp. (NAPS), Boston, MA, USA, 2011, pp. 1–6.

[29] M. Khederzadeh, "Frequency control of microgrids by demand response", CIRED Workshop – Lisbon 29–30 May 2012.

[30] M. Zugno, J. Morales, P. Pinson, and H. Madsen, "Modelling demand response in electricity retail markets as a Stackelberg Game", In International Association for Energy Economics International Conference, Perth, Australia, 2012.

[31] S. A. Pourmousavi, and M. H. Nehrir, "Real-time central demand response for primary frequency regulation in microgrids", *IEEE Trans. Smart Grid*, Vol. 3, No. 4, pp. 1988–1996, December 2012.
[32] PJM, PJM Power Market, online. ⟨http://www.pjm.com⟩; 2013. [accessed 14. June 18].
[33] N. O'Connell, P. Pinson, H. Madsen, and M. O'Malley, "Benefits and challenges of electrical demand response: A critical review", *Renewable and Sustainable Energy Reviews*, Vol. 39, pp. 686–699, 2014.
[34] S. A. Pourmousavi, and M. H. Nehrir, "Introducing dynamics demand response in LFC-Model", *IEEE Trans. Power System*, Vol. 29, Issue. 4, pp. 1562–1572, January 2014.
[35] Q. Zhu, R. Wang, Q. Chen, Y. Liu, and W. Qin, "IOT gateway; Bridging wireless sensor network into Internet of Things", In IEEE IFIP International Confernce on Embedded and Ubiquitous Computing, 2010.
[36] L. Atzori, A. lera, and G. Morabito, "The Internet of Things", *Computer Networks*, Vol. 54, pp. 243–259, October 2010.
[37] W.-L. Chin, W. Li, H.-H. Che, "Energy big data security threats in IoT-based smart grid communications", *IEEE communication Magazine*, Vol. 55, Issue 10, pp. 70–75, 2017.
[38] O. Vermesan, and P. Friess, "Internet of Things strategic research and innovation agenda", In Vermesan, O. , & Friess, P. (Eds.). *Internet of Things: Converging Technologies for Smart Environments and Integrated Ecosystems*, Denmark: River Publishers, pp. 9–30, 2013.
[39] R. K. Bhatia, and V. Bodade, "Smart grid security and privacy: Challenges, literature survey and issues", *International Journal of Advanced Research in Computer Science*, Vol. 4, Issue 1, pp. 1–12, 2014.
[40] H. Mortaji, S. H. Ow, M. Moghavvemi, and H. A. F. Almurib, "Load shedding and smart-direct load control using Internet of Things in smart grid demand response management", *IEEE Transactions on Industry Applications*, Vol. 53, Issue 6, pp. 5155–5163, 2017.
[41] M. H. Yaghmaee, A. Leon-Garcia, and M. Moghaddassian, "On the performance of distributed and cloud-based demand response in smart grid", *IEEE Transactions on Smart Grid*, Vol. 9, Issue 5, pp. 5403–5417, 2017.
[42] T. Brown, S. Newell, K. Spees, and D. Oates, *International Review of Demand Response Mechanisms*. Sydney, Australia: The Brattle Group; 2015.
[43] L. Bird, M. Milligan, and D. Lew, *Integrating Variable Renewable Energy: Challenges and Solutions; National Renewable Energy Laboratory: Golden*, No. NREL/TP-6A20-60451). National Renewable Energy Lab. (NREL), CO, USA, 2013.
[44] M. Farhoodnea, A. Mohamed, H. Zayandehroodi, and H. Shareef, "Power quality impact of grid-connected photovoltaic generation systems in distribution networks", In Proceedings of the 2012 IEEE

Student Conference on Research and Development, Pulau Pinang, Malaysia, 5–6 December 2012.

[45] L. L. Dulau, "Economic analysis of a microgrid", In Proceedings of the International Symposium on Fundamentals of Electrical Engineering, Bucharest, Romania, 28–29 November 2014.

[46] Rocky Mountain Institute, *Demand Response: An Introduction Overview of Programs*. Boulder, CO: Technologies and Lessons Learned; Rocky Mountain Institute, 2006.

[47] S. Annala, Household's Willingness to Engage in Demand Response in the Finnish Retail Electricity Market: An Empirical Study. Lappeenranta University of Technology: Lappeenranta, Finland, 2015.

[48] P. Cappers, A. Mills, C. Goldman, J. Eto, and R. Wiser, *Mass Market Demand Response and Variable Integration Issues: A Scoping Study*. Berkley, CA: Lawrence Berkley National Library; 2011.

[49] SWECO. *Study on the Effective Integration of Distributed Energy Resources for Providing Flexibility to the Electricity System*. Brussels, Belgium: European Commission, 2015.

[50] P. Bertoldi, B. Boza-Kiss, and P. Zancanella, *Demand Response States in EU Member States*. Brussels, Belgium: European Commission, 2016.

[51] H. Pezeshki, A. Arefi, G. Ledwich, and P. Wolfs, "Probabilistic voltage management using OLTC and dSTATCOM in distribution networks", *IEEE Trans. Power Deliv.*, Vol. 33, pp. 570–580, 2018.

[52] J. L. Mathieu, T. Haring, and G. Andersson, Harnessing Residential Loads for Demand Response: Engineering and Economic Considerations. Zurich, Switzerland: Power Systems Laboratory, 2012.

[53] P. Srividya Devi, Dr. R Vijaya Santhi, and Dr. D.V. Pushpalatha, "Introducing LQR-fuzzy technique with dynamic demand response control loop to load frequency control model", *International Federation of Automatic control(IFAC-Elsevier)*, Vol. 49, Issue 1, pp. 567–572, 2016.

[54] P. Srividya Devi, and R. Vijaya Santhi, "Introducing LQR-fuzzy for a dynamic multi area LFC-DR model", *International Journal of Electrical and Computer Engineering (IJECE) IAES Publishers*, Vol. 9, No. 2, pp. 861–874, April 2019.

11

IoT-Based Infant Cradle Monitoring System

M. BHARATHI[1], M. DHARANI[2], T. VENKATA KRISHNAMOORTHY[3], AND D. PRASAD[4]

[1]*Asst. Professor, MB University, Tirupati, Andhra Pradesh, India*
[2]*Associate Professor Sri Vidyanikethan Engineering College, Tiruapti, Andhrapradesh, India*
[3]*Associate Professor Sasi Institute of Technology & Engineering, Tadepalligudem, Andhrapradesh, India*
[4]*Associate Professor Sasi Institute of Technology & Engineering, Tadepalligudem, Andhrapradesh, India*

Contents

DOI: 10.1201/9781003269991-11

11.1 Introduction

Internet of Things (IoT) devices collect data and transfer it to a central data server, where the data are processed, compiled, distilled, and used to facilitate a variety of operations. The applications of IoT are the business community, the government, organisations, and the average user. Mobile phones, computers, coffee makers, refrigerators (mine automatically purchases replacement water filters!), Google Home, Apple watches, and Fitbits are just a few examples of specific IoT-enabled devices. Any device that has sensors and an Internet connection can be connected to the IoT. The public and private sectors have many potential applications for the IoT. People can now track items like their home's security systems, lost pets, and appliance maintenance appointments thanks to the IoT. Customers can utilise the IoT to assist them to make restaurant reservations, keep track of their fitness goals and general health, and receive coupons for a store just by passing by the establishment in question. The IoT can be used by businesses to track supply chains, follow customer spending patterns and feedback, maintain inventory levels, and do proactive maintenance on their machinery and devices. IoT technology naturally benefits blockchain, which is increasingly employed as a more effective and secure way of transaction and data processing. The IoT and blockchain will likely combine more frequently in the future.

11.1.1 IoT Applications

11.1.1.1 Farming-Related Applications of IoT The IoT makes monitoring and managing microclimate conditions for indoor planting a reality, which in turn boosts output. IoT-enabled devices can monitor soil moisture, nutrients, and meteorological information to better manage irrigation and fertiliser systems for outdoor planting. For instance, this prevents resource waste if sprinkler systems only release water when necessary.

11.1.1.2 IoT Applications in Healthcare Wearable IoT devices allow hospitals to keep tabs on their patients' health when they are away from the hospital, thereby shortening hospital stays while still giving up-to-the-second real-time information that could save lives. Smart beds in hospitals keep staff aware of availability, reducing the time spent waiting for a free space. Adding IoT sensors to essential machinery will result in fewer failures and more dependability, which might be the difference between life and death. With the IoT, elderly care is substantially more comfortable. Sensors can tell if a patient has fallen or is having a heart attack in addition to the real-time home monitoring stated above.

11.1.1.3 Applications of IoT for Users IoT devices like wearables and smart homes make life simpler for average people. Accessories like Fitbit, cell phones, Apple watches, and health monitors, to mention a few, fall under the category of wearables. These devices enhance network connectivity, health, and fitness as well as enjoyment.

Environmental controls are activated in smart homes so that your home is at its most comfortable when you arrive home. Dinners that call for an oven or crockpot can be started from a distance so that they will be ready when you get there. Additionally, security is made more accessible because of the user's ability to remotely control lights, appliances, and a smart lock that lets the right people enter the house even when the door is locked.

11.1.1.4 Applications of IoT in Insurance Even the insurance sector can profit from IoT development. Discounts for IoT devices like Fitbit can be provided by insurance firms to their policyholders. By using fitness tracking, the insurer may provide personalised policies and promote healthier lifestyles, which in the long term is advantageous for both the insurer and the user.

11.1.1.5 Industrial IoT Applications Another significant winner in the IoT competition is the manufacturing and industrial automation sectors. RFID and GPS technology can assist a manufacturer in tracking a product from its initial placement on the factory floor to its final placement in the target retailer, or the entire supply chain. These sensors can collect data on the distance travelled, the state of the product, and the environmental factors the goods were exposed to. Sensors affixed to manufacturing machinery can be used to locate production-line bottlenecks, minimising downtime and waste. Other sensors that are put on those same devices can monitor their operation, forecast when they will need maintenance, and stop expensive breakdowns.

11.1.1.6 Blockchain Technology Blockchain is a distributed, unchangeable database that makes it easier to track assets and record transactions in a corporate network. An asset may be physical (such as a home, car, money, or land) or intangible (intellectual property, patents, copyrights, branding). On a blockchain network, practically anything of value may be recorded and traded, lowering risk and increasing efficiency for all parties. Blockchain is a shared ledger that can only be accessed by members of a permissioned network. Blockchain is perfect for distributing such information since it offers immediate, shared, and total transparency. Among other things, a blockchain network can track orders, payments, accounts, and production. Additionally, because everyone has access to the same version of the truth, you can see every aspect of a transaction from beginning to end, increasing your confidence and opening up new prospects.

Parents today battle to really focus on their children since they are so busy. To really focus on and raise their kids, they may need to employ a nanny or quit their jobs. We are mindful of the difficulties parents experience while attempting to raise their children, especially when the two parents are busy. Living 24 hours in such conditions is exceedingly difficult. We actually wanted to make a supportive strategy to resolve the issue utilising IoT innovations. The physical and computerised worlds are connected by the IoT. The Internet and other smart machines and devices are connected. They accumulate relevant data about the area around them, examine it, and connect it. Information from the sensors is sent to monitoring devices like Arduino and NodeMCU, permitting the devices to perform explicit assignments using IoT innovation, which is planned to diminish business-related blunders and save time. We will make a support framework with the assistance of IoT that will decrease parents' feelings of anxiety and, in particular, be completely safe for the infant regarding both time and security. Hence, using time effectively and child care are two pivotal issues.

In addition, in the event that the child has a cold or fever, cradle frameworks are perfect for spotting it and conveying the message. For the child's security, our support can likewise recognise any movement utilising a movement sensor. The proposed approach will help parents in raising their children well. As we have found in India and other industrialising nations, the two parents may both work and take care of the baby, which expands their responsibility and may affect both their day-to-day work schedules and the lives of their children. We are building a cutting-edge support framework because of parents' busy schedules andsmaller support systems.

We are very mindful of the challenges parents experience while attempting to raise their children, especially when the two parents work. Living in such circumstances 24 hours [1] a day is basically unworkable. Thus, we should make something extraordinary that will allow parents to monitor their child or baby continually and will advise them regarding any changes.

To empower parents to monitor their children even while they are away from home and to have the option to identify the behaviour of the child from anyplace on the planet, we have concocted the idea to make a smart support framework utilising IoT.

The IT business is fostering an enormous number of IoT devices [2]. Although a few supports are developed with IoT incorporation, there are still a few components that could threaten the infant's wellbeing (Figure 11.1).

11.2 Literature Survey

The framework that joins Hadoop and the C4.5 calculation to gauge problems utilising the accumulated data is shrouded in [3] a smart and secure IoT-based child behaviour and health monitoring system. [2,4] "IoT-Based Healthy Infant Cradle System," investigates a framework that utilises the IoT, Amazon Web Service, and Smart Baby Cradle, and offers parents a brilliant framework to help them in checking and soothing the child. [1] "A Deep Learning-Based IoT-Based Intelligent Baby Care System" utilises a childcare framework and the IoT

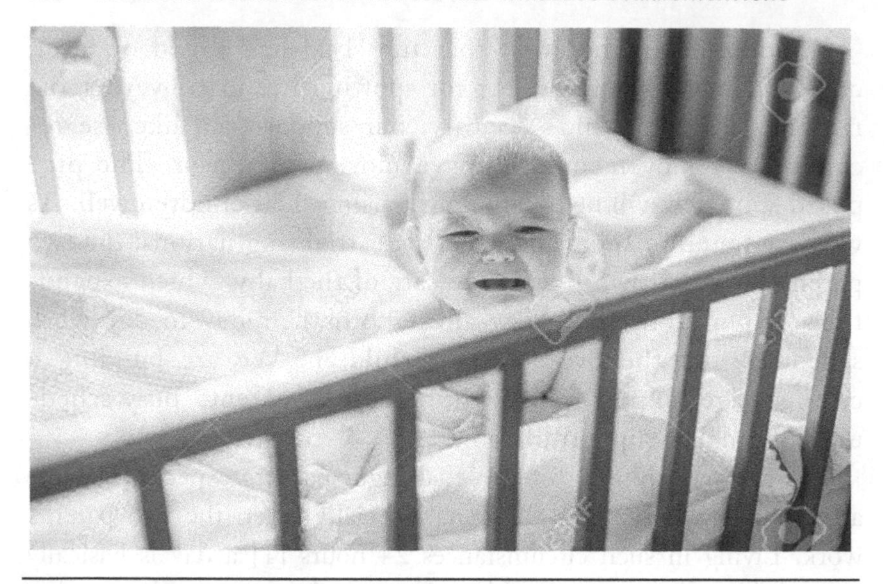

Figure 11.1 Challenges to parents to bring up their kid.

to assist with child care. Smart learning methods are likewise utilised. Through smart learning, this device monitors the child's conditions, including position, internal heat level, and stance, empowering parents to monitor the condition of their child.

Rather than a parental figure, a support framework is used, which mitigates the baby by playing music and talking with them. It watches out for the newborn child and gives reports on their well-being and mental state.

11.3 Proposed System

The observed measurements, like temperature and sound, are remembered for this recommended IoT framework. The fan [5] will turn ON when the room temperature is high. The DC engine will pivot, and a melody will be played through the speaker [6] to make it bedtime for the child in the event that the sound sensor [7] perceives the child crying. Here, we're exhibiting support development with a DC engine. Temperature and humidity values are sent to a cloud server (Figure 11.2).

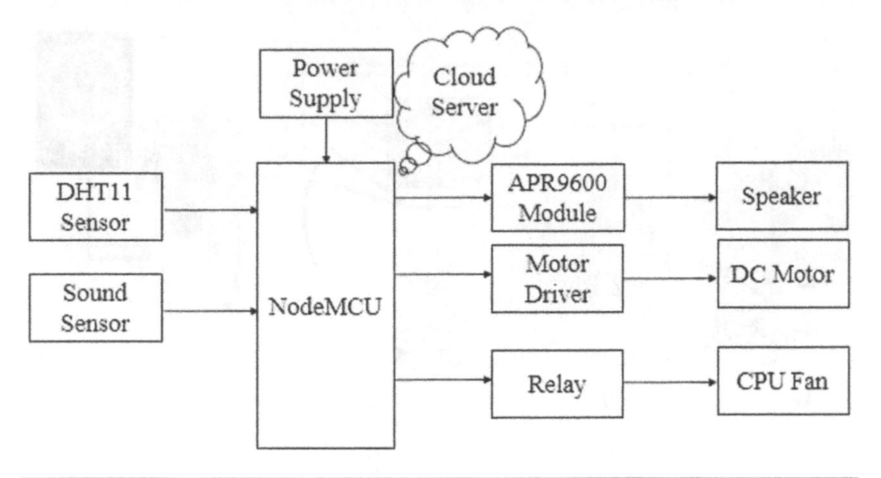

Figure 11.2 Block diagram of proposed method.

With this framework, parents can monitor the child's immediate environment, including the temperature and humidity, by means of a cell phone and the Thingspeak site. The temperature and humidity readings are kept on a cloud server. This device enacts the fan when the temperature surpasses a predetermined value. Assuming the newborn child cried, the support begins to move and music begins to play to calm the child.

This strategy is positioned to keep the baby agreeable for any reason, for example, when the parents watch the baby because they need to get work done. The child feels uncomfortable at certain times; subsequently, this equipment is useful because it makes dealing with the child easier. As it were, it's useful that when the child cries, the support moves and music begins to play to attempt to stop the crying. This innovation enacts the fan to bring down the temperature and humidity around the child when the temperature surpasses a predefined limit.

11.3.1 Working Principle

The child monitoring cradle system will look after the baby and, in the event that the parents are not there, will make them aware of the baby's environmental factors.

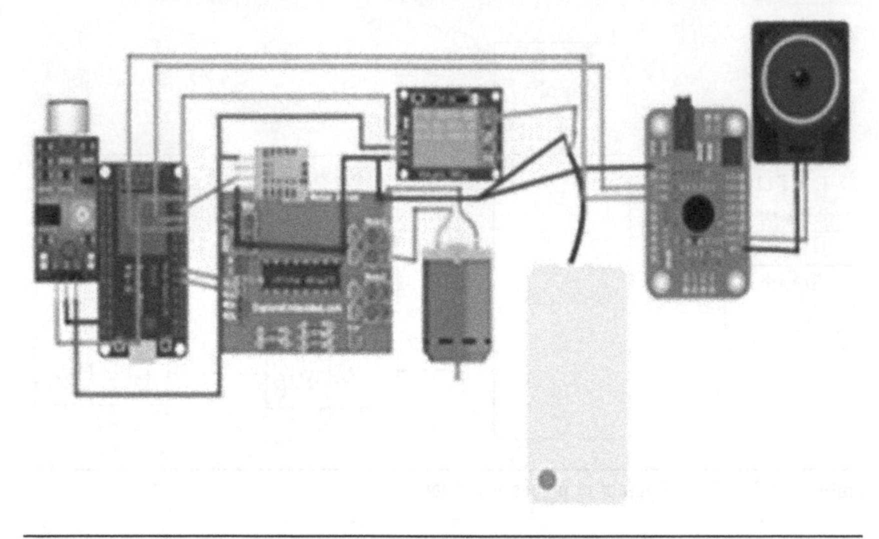

Figure 11.3 Working model developed in proteus environment.

In this circuit, the power supply module supports moving a consistent DC supply with next to no variances to the circuit.

The engine driver module, node MCU [8], and relay [9] have all been connected to the power supply module [10]. In this engine driver, the sign to enact the DC engine, which swings the support because of the child's shouts, is utilised (Figure 11.3).

Soothing music should be kept on the module. Also, the power supply module gives power to the module. The baby crying is received by the sound sensor, which conveys the message to the NodeMCU. With the assistance of the APR9600 module [11], a melody will play when the child cries, giving them the feeling that the parent is with them.

Since these are electrical devices and, additionally, the framework is held near the child, parents should practice the highest level of caution in preventing flow spillage. The relay is utilised to stop the opposite stream or spillage of current to prevent any harm.

The DHT11 sensor distinguishes the temperature and humidity levels around the newborn child, and in the event that the numbers it measures surpass a specific limit, a fan enacts to adjust the surrounding environment. Utilising the Thingspeak cloud server on their cell phone, parents can monitor the climate their kid is in (Figure 11.4).

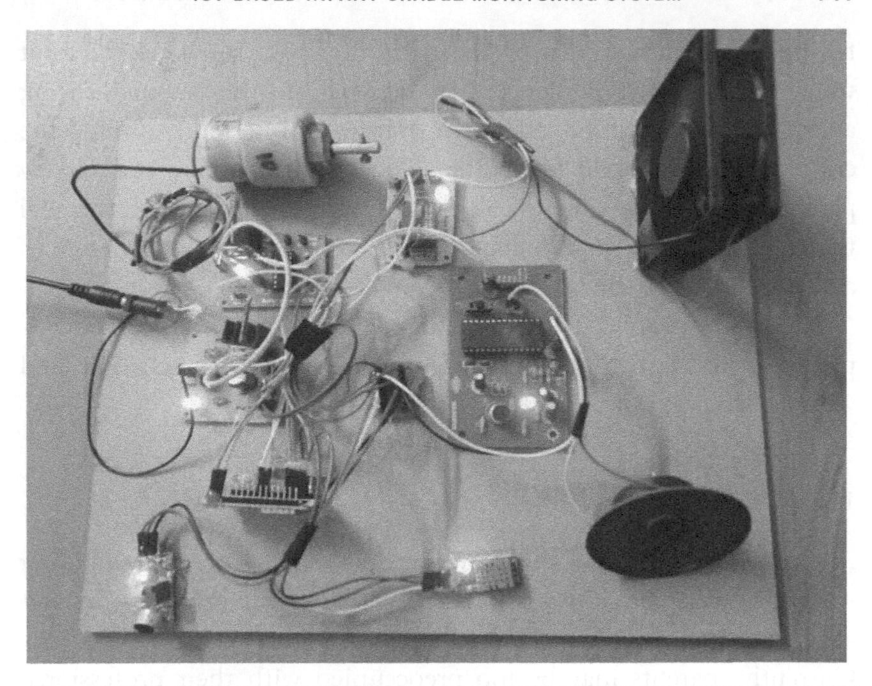

Figure 11.4 Experimental setup and implementation.

The infant cradle monitoring system using IoT is connected to the power supply using the circuit shown above.

11.4 Results and Discussion

Thus, a cradle monitoring system using IoT was made to keep the baby agreeable under all conditions, for example, when parents are too preoccupied with their responsibilities to invest their energy in the baby. The child feels uncomfortable at certain times; hence, this equipment is useful in light of the fact that it makes dealing with the child more straightforward.

The framework, which is comprised of the Node MCU, APR9600 module, DC engine, and DHT11 sensor, has capabilities such that, because of a child's cries, tells the APR9600 module to play a tune that will calm the child and trains the Node MCU to teach the Motor_driver [12,13] to turn in both clockwise and counterclockwise headings for a predetermined timeframe.

This framework additionally recognises the temperature and humidity of the air around the child, and when the gathered value of temperature and humidity by the DHT11 sensor surpasses the limit value, the Node MCU conveys the message to the fan to turn on, assisting with diminishing the temperature and humidity around the child to help the child feel better. The qualities can be checked by parents utilising the Thingspeak cloud server whenever, with the goal that they will work in hazardous circumstances, and they are tested in an air conditioned environment.

Thus, the support checking framework utilising IoT empowers parents to deal with their children even with a busy schedule and without stressing over the little details.

11.5 Conclusion

A new generation of healthcare monitoring systems combining wearable electronics and photonics is now possible because of recent advancements in sensor and wireless communication technology. Currently, parents may be too preoccupied with their professional lives to have enough time to care for their infants. A nanny can be too pricey for the family to pay for. Today's parents must juggle their home and job obligations concurrently. They have to look after the house and the kids after a long day at work. They might not have enough time to manually swing their cradle and comfort the infant.

Furthermore, given the way that people live today, it is quite challenging for even housewives to sit next to their babies and comfort them whenever they scream. This device can enhance parenting because it is affordable and easy to use. Parents feel more secure as a result. The continuous monitoring of a baby's many biological indicators and analysis of their general health enables a mother to comprehend the general state of the baby's health and, if necessary, to take appropriate action. Also measured is the cradle temperature. When a baby cries, the music can calm them while also alerting the parents to the issue.

References

[1] B. Mohammad, H. Elgabra, R. Ashour, and H. Saleh, "Portable Wireless Biomedical Temperature Monitoring System", IEEE International Conference Publication on Innovations in Information Technology (IIT), 19 March 2013.

[2] E. Saadatian, S. P. Iyer, C. Lihui, O. N. N. Fernando, N. Hideaki, A. D. Cheok, Z. Amin (2011). "Low Cost Infant Monitoring and Communication System", IEEE International Conference Publication, Science and Engineering Research, 5–6 December 2011.
[3] R. Nicc, N. P. Jain, P. N. Jain, and T. P. Agarkar, "An Embedded, GSM Based, Multiparameter, Realtime Patient Monitoring System and Control", IEEE Conference Publication in World Congress on Information an Communication Technologies, 2 November 2013.
[4] M. Bharathi, D. L. Rani, N. Padmaja, and M. Dharani, A Health Monitoring System Based on IOT for Persons in Quarantine. *Journal of Algebraic Statistics*, 13(3), 387–392, 2022.
[5] https://en.m.wikipedia.org/wiki/Computer_fan
[6] http://www.schoolchalao.com/basic-education/show-results/electronic-equipment/computer-speaker
[7] https://robu.in/product-category/sensor/sound-sensor/#:~:text=A%20sound%20sensor%20is%20defined,that's%20highly%20sensitive%20to%20sound edition, Prantice-Hall of India, New Delhi, 2002.
[8] https://www.electronicwings.com/nodemcu/introduction-to-nodemcu
[9] https://www.electronicshub.org/what-is-relay-and-how-it-works/
[10] https://www.techopedia.com/definition/1756/power-supply
[11] https://www.digchip.com/datasheets/parts/datasheet/811/APR9600_amp.php
[12] https://2020.robotix.in/tutorial/auto/motor_driver/
[13] https://byjus.com/physics/dc-motor/

Index

Note: *Italicized* and **bold** page numbers refer to figures and tables.